JN147025

The Theory of Stability
安定性理論

赤沼 篤夫

東京図書出版

Contents — 目次

The Theory of Stability 3

安定性理論 59

The Theory of Stability

Prologue

The conceptual existences in the human mind are stable and are substantially perceived with scholar values. But physical existences are independent of the human conception and have some instability. They are subject to the natural regulations which are essentially contingent. Individual physical events and existences are substantial products from the contingency and are controlled by the regulations induced by the contingency. Nature is actualized by the stabilization which the contingency brings forth and events follow the stabilization processes. The statistical perception on the essentialities of natural events induces this theory of stability. This concept on the theory of stability consists of the following two principles.

1. Principle of insecurity; Events exist. Existence involves some statistical insecurity.
2. Principle of inconstancy; Events persist. Persistence involves some statistical inconstancy.

When the existing situation changes, events vary and stabilize for continuity. Events have variability and continuity. For the variability, the probabilities of existence and persistence are proportional and for the continuity, the probabilities of existence and persistence are inversely proportional. Variability and continuity substantialize events. Events evolve and stabilize. Events with low continuity will change into high continuity reducing variability with favorable conditions to exist easier.

These concepts stated above explain that the theory of stability is consistent with the fundamental physical laws. The universal gravity is derived from the principle of insecurity and the theory of special relativity is derived from the principle of inconstancy. These principles deduce various natural propositions including variability and continuity, which explain existence and evolution of events. The principles are expected to be applied to various events and also used in various studies. The concept of variability and continuity should be acknowledged and is applicable to various subjects.

Contents

Prologue 5

A. The essentiality of existence 11

1. The concepts of space, time and energy 11
   (1) Space   (2) Time   (3) Energy

2. Existential insecurity and inconstancy 13

B. Existential insecurity and gravity 15

1. Principle of insecurity 15

2. Existential insecurity 15
   (1) Dispersion   (2) Existence on a point

3. Existential probability distribution 16
   (1) Probability distribution   (2) Restoration force

4. Displacement and centripetal force 18
   (1) Precision and gravity   (2) Occurrence and motion

C. Existential inconstancy and relativity 20

1. Principle of inconstancy 20

2. Existential inconstancy 20
   (1) Persistence   (2) Fluctuation

3. Persistence probability distribution 21
   (1) Probability distribution   (2) Momentum

4. Velocity and relativity 24

D. Hypothetic existence 26
1. Hypothetic state of energy 26
2. Hypostatic forces 26
3. Substantial existence and hypostatic state 27

E. Actuality of a mass point 28
1. Variability of a mass point 28
2. Continuity of a mass point 28
3. Substantiality of a mass point 29
4. Uncertainty of a mass point 29

F. Actuality of two mass points 32
1. Linear and spatial occurrence 32
2. Occurrence of another mass point 32
3. Influence and rotation of another mass point 34
4. Rotation 36

G. Basic propositions 37
1. The universal gravity 37
2. Coulomb force 38
3. Mass and energy 39
4. Stabilizing events and revolution 40
   (1) Substantiality (2) Variability (3) Continuity (4) Revolution
5. Uncertainty of events 45

    (1) Uncertainty principle   (2) Variability of momentum
    (3) Uncertainty and variability

  6. Characteristics of heat and light 47

  7. Variability and Lorenz function 49

H. Special relativity and Lorenz function 50

  1. Relativity of motion 50

  2. Relativity of length 51

  3. Relativity of time 52

  4. Coordinate transformation 54

    Epilogue 56

A. The essentiality of existence

1. The concepts of space, time and energy

(1) Space

It is the mental perception in physics what space actually is. Whatever is the space of great universe, the space of our vicinity has been percepted as the three-dimension space. Therefore, the existence of objects is usually considered in the three-dimension space which is Euclidean space. Space which is an aggregate of points is determined by an origin and three fundamental directions which is one Euclidean space. Several other Euclidean spaces can be considered to exist in one Euclidean space.

A point in a Euclidean space is considered to have position and motion. The position of a point has the relativity to another point and the motion is the change of relativity. A line is a range of points. Points, lines and shapes are defined as a point or an aggregation of points and, therefore, can be expressed in a Euclidean space. A conceptual point in a shape coincides perfectly with a point in a Euclidean space, for example, to the origin with conceptually perfect precision and with no instability. A conceptual point changes its position, then its relativity to the origin changes. This change is called motion. These conceptual points and motions are employed in the actual perception of points and motions which should have in nature some instability.

(2) Time

Time is considered as an element of an event in the space. Time is an independent function of existence. A static or moving mass in a space has an independent value of time whereas other elements in a space, for example, a distance in a fundamental direction can become a dependent value of time. Hence space and time construct a frame of existence. A space-time frame is required to express the physical existence of an event. In Euclidean spaces time is fixed to all spaces and has no relativity of time, which is Galilean relativity. Every space has to have its own time and to have the relativity to each other.

Time is a component of energy. Space and time are required to express the existence of a physical event. The existence of a mass point can be expressed in a space-time frame. And a space-time frame can also have various space-time frames in it. Each space-time frame has its own values of spatial elements and time. When there is no change or no motion of each frame in a reference frame or each frame is fixed to the reference frame, then time is also fixed to all frames and has no relativity of time among frames. The relativity of time occurs to dynamic events or dynamic frames.

(3) Energy

An essential existence is energy. The physical existence involves time space and energy. Energy has intensity and quantity. Energy is an existence. It should have position, motion and existential distribution. The existence of energy should have the statistical exactitude or reliability, which means energy also has the existential insecurity and inconstancy. Hence the presence of energy follows the principles of insecurity and inconstancy. But its inconstancy is large, therefore, its variability is also large. It easily changes. High-intensity energy forms light quanta and mass points. Light quanta and mass points are energy carriers. The carrier has position and motion and their own substantiality. But light quantum has almost no inconstancy, therefore, it has essentially no substantiality. The higher intensity energy forms substance or mass. Their substantiality is comprehended here using mass points.

When the point is not conceptual but physically actual, which is a mass point, it does not always coincide perfectly with the conceptual point. It can shear from the point. The coincidence is mental and is essentially not perfect and its exactitude is statistical. The coincidence is not secured, which is substantiality.

A. The essentiality of existence

Shears or deviations accompany the coincidence. Shear or deviation decreases the statistical exactitude or reliability, which is existential insecurity. The expected position which is usually the center of existence probability distribution has the higher reliability. When a physical point would not be in the center, it would move to the position with higher reliability which resides at the center. The reliability difference causes the centripetal force and is the source of gravity.

A solid object, which is solidified energy or substance, is also considered as an aggregation of physical points, therefore, its presence can be expressed in a space-time frame. The presence of a physical object in a space-time frame does not necessarily coincide to the human conceptual presence. The human thought or conception is a scale and its quantity is the scholar. The physic-quantity measured with this scale is the statistic. An aggregation of points with insecurity in this meaning is a physical object or substance.

Physical objects have position and motion, and they have relative difficulties in the persistence of motion. They are called mass. Mass is designated to the relative difficulty in changing the motion velocity depending on its size and sort. Essentially inconstancy probability represents the difficulty difference. A physical object is often represented by one geometrical point neglecting its shape and size. The geometrical point represents the mass of an object, which is the mass point. Presence of energy is often represented by one geometrical point and the physical quantity of energy is given to this point neglecting its shape and size which are hardly possible to be grasped.

2. Existential insecurity and inconstancy

A point which a man thinks about is a conceptual point which can be placed exactly at a point in a Euclidean space, and can make free conceptual movement. The point stays at a point in a Euclidean space with perfect accuracy and its movement has complete steadiness. A substantial point based on nature, which is a mass point and can be considered as the simplest existence, cannot stay or move exactly as a man thinks about. It is an independent existence from human conception and follows natural laws. Conceptual existence can be artificial but the physical existence which follows natural laws is always involved with contingency. A point of an object exists with certain insecurity and is independent of human thought. There is the essential difference between the existence of the conceptual point and the existence of a point of a physical object in a Euclidean space which has shears and shakes. A mass point has position and motion. The position can comprise gaps and deviation, which is designated insecurity. And motion can have fluctuation, which is designated inconstancy.

B. Existential insecurity and gravity

1. Principle of insecurity

Events exist. Existence is statistic quantity and involves some statistical instability which is insecurity.

When a mass point is expected at point $x = a$, its deviation (X) is $(x-a)$. A point at deviation (X) has the existential probability which is expressed with normal function (Φ),

$$\Phi = \exp[-X^2/(2\sigma^2)]/[\sigma(2\pi)^{1/2}]$$

Existential precision is not perfect. Existence always comprises some insecurity which is expressed with standard deviation (σ).

2. Existential insecurity

(1) Dispersion

The sensuous evaluation of existence is made here. A man places, for a practical example, a golf ball on a flat surface. A set of coordinates is made on the surface. The gravity center point of a golf ball is placed at a specific point, for example, at the origin. The gravity center can be placed at the origin, which is although conceptual. Errors are always involved in the reality. When this trial is repeated 100 times, it is difficult to think that the center point is always placed on origin. There should be dispersion. If college students perform this trial, the modal point may be the origin and the dispersion may be relatively small. If primary school boys perform this trial, the modal point may also be the origin. But the dispersion becomes larger. In both trials, the gravity center point is placed at the origin but the nature of existence is different. The location is represented by the origin but the dispersion accompanies with various standard deviation.

The former trial has the smaller mean deviation and the higher precision where the locations are gathered in the smaller vicinity. The latter trial has the lower precision with the larger errors and the lower security. There is the difference in the existential precision and the latter trial has the larger scattering and the larger standard deviation value.

(2) Existence on a point

The existence is evaluated from the other direction. The gravity center of a golf ball is located at origin. Let the center represent the golf ball and is considered as the mass point which exists at the origin. Conceptually the mass point is exactly secured at the origin. The mass point exists very likely at the origin but there some insecurity involves. There could be dispersion. There could involve some errors. The reliability of the existential probability distribution of the mass point is the maximum at no dispersion or at the expected location which is the origin point and around it there lie lower probability densities. This situation is involved with a high or low precision of the existence and is designated insecurity.

This situation is easily conceivable with particle locations. When the position of electrons or other particles in one instance is considered, the position can be calculated. But not all particles are at the calculated position. The value comprises some statistical dispersion which is insecurity. The calculated position is the modal value. Some particles are scattered around it. It is statistical existence.

3. Existential probability distribution

(1) Probability distribution

Existential probability distribution indicates the possible existence at the expected point and its surroundings. The deviation from the expected point is the stochastic variable. The expected position is the modal value. It is the mean value

B. Existential insecurity and gravity

of the variable in the distribution which is a normal distribution. The reliability of the modal value is the existential precision. The higher the value of this reliability is, the more stable the existence is. The value of standard deviation defines insecurity. The smaller the value of the standard deviation is, the more stable the existence is. The precision is defined by the reciprocal of standard deviation. To make it simple, the distribution of the x direction is considered. The occurrence on x-axis is all over the same. Hence the probability distribution is the normal distribution (Φ) with a standard deviation (σ) and the expected point which is ($x = a$) in the center.

(1) $\quad \Phi = \exp[-(x-a)^2/(2\sigma^2)]/[\sigma(2\pi)^{1/2}]$

σ is the standard deviation of $\Phi \quad \sigma > 0$

X shall be deviation on x-axis from the expected point (a), then $X = x - a$.

(2) $\quad \Phi = \exp[-X^2/(2\sigma^2)]/[\sigma(2\pi)^{1/2}]$

Reliability of the expected point ($X = 0$) is $1/[\sigma(2\pi)^{1/2}]$. The higher this value is, the steeper the reliability decreases, therefore, small gaps from (a) hardly exist with high reliability, which is precise existence. Reliability decreases as standard deviation (σ) increases, therefore, the magnitude of (σ) represents insecurity and the reciprocal ($1/\sigma$) expresses existential precision.

(2) Restoration force

Suppose a mass point now stays still at the origin, and if the standard deviation is small, which is high value of precision, then the mass point has a high reliability and makes little deviation. A small displacement makes the abrupt decrease of the reliability. Then a large difference of the reliability occurs and the mass point has high possibility to recover the position right away. That means force works toward the expected point. When the mass point would have lower reliability, then a small displacement would slightly decrease the reliability, therefore, the restoration force is small. The existence of low insecurity means that the restoration force is relatively high and the mass point has the higher probability to stay at the expected point. This condition has high precision and small existential insecurity. The restoration force which is the centripetal force at the expected point is designated as the central force.

4. Displacement and centripetal force

(1) Precision and gravity

A stationary mass point has uniform centripetal force around itself. Therefore, it can keep the static condition. A small displacement decreases the existential reliability and the restoration force works since the center has higher existential reliability. The centripetal force is the strongest at the expected point and it decreases proportionally to reliability in the peripheral. Existence with high precision has a strong centripetal force and hardly makes a deviation. Existence with low precision possibly makes deviation since central force is not very strong. It is proportional to reliability at the center, which is the expected existential point. And it is proportional to the existential precision ($1/\sigma$). From equation (2) the central force (F) is proportional to $[1/\sigma(2\pi)^{1/2}]$.

$$(3) \qquad F = E/[\sigma(2\pi)^{1/2}]$$

Proportional constant E has the dimension of energy and is considered as the mass point energy. The centripetal force, which causes the gravity, is a function of X. If there exist energy K at deviation X, the reliability of energy E causes a centripetal force to energy K toward energy E. The centripetal force is the reliability at X multiplied by energy K. At the same time, energy K has the existential reliability at $X = 0$ which causes a centripetal force to energy E toward K. The product of both centripetal forces makes the gravity.

(2) Occurrence and motion

Let E be the energy at the expected point and let P_C be the occurrence of a unit of energy. Then the relationship between standard deviation (σ) of existence probability distribution and the occurrence (P_C) of unit of energy is the following,

$$(4) \qquad \sigma^2 = P_C(1-P_C)/E$$

P_C is small enough compared to 1,

$$(5) \qquad \sigma^2 \fallingdotseq P_C/E$$

The standard deviation is the square root of the quotient of occurrence and energy. Mean variance is the same as the occurrence of a unit of energy or the occurrence rate. Hence the square root of energy and the reliability of existence probability are proportional. And the reliability and the square root of occurrence are inversely proportional. A mass point in the space of uniform occurrence has all the same mean variance of existence probability at any point in the space. A mass point in that space is stable and has no motion. But the reliability of existence probability is not 100%. The possibility to change its position coexists. Should it change the position, the existential reliability does not change. The same centripetal force works. The centripetal force around the mass point is uniform and the mass point exists stable. If occurrence would be non-uniform in the space and point (a) in the space would have the lower occurrence which is the higher reliability of existence probability. The point (a) has the higher existential reliability and the centripetal force works toward the point (a). The mass point staying at point (b) which is the deviation X of point (a) moves toward point (a). The mass point obtains the higher existential reliability and the higher stability.

C. Existential inconstancy and relativity

1. Principle of inconstancy

Events persist. Persistence is statistic quantity and involves some statistical instability which is inconstancy.

Persistence has the inconstancy probability with fluctuation (V), which is expressed with normal function (Ψ).

$$\Psi = \exp[-V^2/(2\tau^2)]/[\tau(2\pi)^{1/2}]$$

Persistence is not perfect. Persistence always comprises some inconstancy which is expressed with standard deviation (τ).

2. Existential inconstancy

(1) Persistence

Among the existences of the same precision, some mass points change their positions easily and some hardly change. Sensuous consideration on quivering is made here. A ping pong ball shall be placed at the origin of a set of coordinates. College students perform this action 100 times. The precision was as high as in the action with golf ball. But the stability is not as high as with golf ball. Ping pong ball is easy to move, that means it is weakly fixed. A golf ball stays still better. A golf ball has the better impulse for standstill. A golf ball has high probability to have persistence. The existence with the same precision can have different steadiness or persistence. Now the weight center of a golf ball exists still at the origin point, which means there is no quivering or fluctuation speed is 0. That is the existence in which the steadfast situation has the maximum persistence. The higher the reliability is, the stronger fixture to the position is. The persistence possibility of a ping pong ball is much less compared to a golf ball and its steadfast situation has rather less fixture to the position.

(2) Fluctuation

When the existence of a point is considered, it has not only displacement possibility but also fluctuation possibility. An existence has essentially two elements of instability which are existential insecurity and inconstancy. Existence has not only positional instability whether a mass point is on the expected point, which is insecurity, but also qualitative instability of persistent motion. The latter instability is defined here inconstancy. Steady motion with high persistence has little motion fluctuation. A steadier golf ball has less fluctuation possibility or lower inconstancy.

There are existences which are easy to fluctuate with high inconstancy or those which are hard to fluctuate with low inconstancy. For example, a heavy ball is an existence with low inconstancy and has relatively strong persistence and therefore, hardly fluctuates. A ping pong ball has low persistence and easily fluctuates. This difference is expressed by the difference of inconstancy. The low inconstancy means that the mass point has a strong fixture to the position, which is high persistence. The same discussion can be made to a shape which is consisted of points. In the strong shape, which is an existence of low inconstancy, the positional relationship of representative plural points on the shape is hard to change.

Existence generally implies the two elements of precision and consistency, which complies that a mass point at a point has not only deviation but also fluctuation.

3. Persistence probability distribution

(1) Probability distribution

A mass point has two elements which are position and motion. Motion instability distributes with normal distribution centering no fluctuation. Its standard deviation expresses the magnitude of the inconstancy. The stochastic variable of this persistency probability distribution is the time dependent change rate of the stochastic variable of existence probability distribution. This change rate is designated here fluctuation velocity or fluctuation. A standstill mass point has fluctuation modal value of 0 and its reliability is the standstill possibility. The probability distribution is a normal distribution and the smaller its standard deviation is, the higher the standstill possibility is, and the more persistent. Hence the reciprocal of the standard deviation is designated here persistency. The situation is considered here where the motion is standstill. When a mass point is standstill at a point (a) in a Euclidean space, then the motion velocity is 0. There is possible fluctuation even in the standstill situation. At a point in a uniform space a mass stays still, then the mass point has uniform fluctuation possibility around it at any direction and at any velocity. In consequence, its mean fluctuation is 0.

Let X be the deviation of a mass point staying at point (a) on x-axis, then the fluctuation velocity (V) is as follows.

(6) $\qquad V = dX/dt$

Fluctuation is the time dependent change rate of deviation and has the dimension of velocity. The mass point has persistence probability of normal distribution centering 0 at any direction. Let (Ψ) be the persistence probability distribution, then its standard deviation (τ) expresses the inconstancy and its reciprocal is the consistency or strength of persistence. Considering on x-axis, a mass point staying still at ($x = a$) has a normal distribution of inconstancy centering 0 of fluctuation velocity.

(7) $\qquad \Psi = \exp[-V^2/(2\tau^2)]/[\tau(2\pi)^{1/2}]$

Reliability at $V = 0$ is $1/[\tau(2\pi)^{1/2}]$, which is the steadfast reliability. The higher this value is, the more abrupt fall of reliability with a small shift makes.

(2) Momentum

A mass point is at a point, which means it stays still and is stable at the point. Its motion velocity is 0. This velocity 0 has some instability. The mass point can fluctuate but it has the maximum reliability at fluctuation velocity 0 in its persistence probability distribution and the lower reliabilities exist around it. If it does deviate from the center, the reliability difference of a persistence distribution makes the motion of mass point to return quickly to the center where the existence has little inconstancy and large persistency. It is, therefore, steady state. The standard deviation τ expresses the magnitude of inconstancy. The smaller it is, the stronger the existential persistency is. The reciprocal of this standard deviation which is $(1/\tau)$ is proportional to the existential persistency. Hence it is designated consistency and is proportional to momentum.

Now a mass point of energy (E_1) stays still ($v = 0$) at point (a) on the x-axis, then the persistence reliability is $1/[\tau(2\pi)^{1/2}]$ applying equation (7). The product of energy E_1 of a mass point and persistency reliability has the dimension of momentum.

$$(8) \qquad P = E_1/[\tau(2\pi)^{1/2}]$$

Hence momentum (P) is inversely proportional to the inconstancy of which proportional constant has the dimension of energy and is considered the energy of mass point.

Product of momentum and angular velocity is force. The quotient of central force (F) and static momentum (P) is angular velocity (ω). F/P is obtained from equations (3) and (8).

$$F/P = [E/(2\pi\sigma^2)^{1/2}]/[E_1/(2\pi\tau^2)^{1/2}]$$

$$= (\tau/\sigma)/(E_1/E)$$

(F/P) is (ω) and (τ/σ) is (ω), therefore, $E_1/E = 1$. E is communal to insecurity and to inconstancy, which is the energy and represents the mass point and defines its

The Theory of Stability

central force and momentum.

4. Velocity and relativity

A mass point should move to a point where the reliability is higher than that at the present position. If motion would not affect inconstancy, the mean fluctuation of a moving mass point which has the dimension of velocity has no change in the persistence probability distribution. But the fluctuation toward the moving direction can change. Let (τ') be the standard deviation of the moving mass in its persistency distribution in the moving direction. A moving mass point can be affected by its velocity, which disturbs the fluctuation. It changes the inconstancy (τ') which can be a dependent function of v. The mean value of the fluctuation in moving direction should be expressed as the product of the velocity and a positive function (f). If there would be no effect from the velocity, function (f) should be 0. The square of τ' is the mean of variation $(V-fv)^2$ which is the subtraction of square of the postulated mean inconstancy $(fv)^2$ from the square of standstill inconstancy which is (τ^2) of equation (7) when approximate fv is close to τ.

$$\tau'^2 \fallingdotseq \tau^2 - (fv)^2$$

When the mass point is standstill ($v=0$), τ'^2 equals to τ^2. As v is extrapolated upward, $(fv)^2$ increases, then τ'^2 decreases. But $\tau'^2 \geq 0$ or more. Hence the maximum value of $(fv)^2$ is τ^2. Let C be the maximum value of v. Then, the following equations are obtained,

$$\tau^2 = (fC)^2$$

$$f^2 = \tau^2/C^2$$

(9) $\qquad (\tau')^2 = \tau^2 [1-(v/C)^2]$

The velocity of a mass point has the limitation which should be the light speed or the speed close to it. It has been considered as the speed of light in the theory of special relativity by Einstein.

The momentum of a mass point in motion is obtained. From equation (8), the momentum in standstill (P_0) is $E/[\tau(2\pi)^{1/2}]$, and the momentum (P) in motion with velocity (v) is $E/[\tau'(2\pi)^{1/2}]$, then equation (10) is obtained,

$$(10) \quad P = P_0/[1-(v/C)^2]^{1/2}$$

As momentum increases, the velocity increases and comes to close to the limitation which has been considered the light speed, but can be more since light is particulated.

D. Hypothetic existence

1. Hypothetic state of energy

The standard deviation in equation (2) defines insecurity. As the standard deviation increases, the existence reliability of a mass point decreases and the substantiality of a mass point disappears with infinite value of insecurity. Equation (3) shows that the centripetal force disappears with infinite insecurity. This state is hypothetic. A stable flat distribution of the existence reliability with positive or negative infinite insecurity is conceivable. A mass point becomes a flat density of existence probability which is represented by the distribution of energy E of equation (3). Hypothetic existence with a negative insecurity is also conceivable where centrifugal force occurs with equation (3) and the energy E represents the existence. And hypothetic energy excess (E) and energy deficits (−E) with negative insecurities can be considered in the hypothetic state. The centrifugal force occurs to an energy excess and the centripetal force to an energy deficit with hypothetic negative insecurity.

2. Hypostatic forces

Equation (2) is applied to the hypothetic mass with negative reliability. X shall be deviation from the expected point of a hypothetic energy mass which is an energy excess or energy deficit.

$$(2) \qquad \Phi = \exp[-X^2/(2\sigma^2)]/[\sigma(2\pi)^{1/2}]$$

The centrifugal force works to an energy excess. And the smaller absolute value of the standard deviation is, the stronger the centrifugal force is. The energy deficit draws energy and the deficits have the tendency to restore. The excess has the tendency to decay. When equation (3) is applied to the excess where F is negative and is centrifugal force. Frailty of the excess at the expected point (X = 0) is $1/[\sigma(2\pi)^{1/2}]$. The higher this absolute value is, the frailer the excess is, therefore, the excess has the more pressure to be flattened. Frailty increases as the absolute number of standard deviation (σ) decreases, therefore, the reciprocal ($1/\sigma$) expresses the tension to decay. The centrifugal force is proportional to the tension ($1/\sigma$). From equation (2) the central force (F) is proportional to $\{1/[\sigma(2\pi)^{1/2}]\}$.

(3) $\qquad F = E/\{1/[\sigma(2\pi)^{1/2}]\}$

Proportional constant E is considered as the excess energy. When E is a unit of energy, the probability distribution is the same as the excess energy incidence at deviation X. $-F$ is the tension in the center of a distribution.

3. Substantial existence and hypostatic state

A particle is a substantial existence with mass which is consisted of energy. An example is a neutron which is a substantial existence. A neutron can become a proton and an electron which are consisted of energy. When a neutron releases an electron, the new existence which is a proton can't always receive perfectly balanced energy. An energy excess which is a hypothetic existence can be left in a new proton. At the same time, the electron has an energy deficit which is also hypothetic existence. These proton and electron generate Coulomb force and the electron exists rotating around the proton with a certain angular velocity. High intensity energy which is the energy of high frequency wave quantize or particulates. Particles have their own proper energy levels and energy excess or energy deficit can happen. These excess or deficit are hypostatic state of energy.

E. Actuality of a mass point

1. Variability of a mass point

A conceptual point can move free in a Euclidean space. But physical points or mass points cannot move freely in the space. The existence probability restricts the motions. The motion is the time dependent relativity change of each point on a shape to the origin of a Euclidean space. Considering on motion, the concept of duration time is required. The elapsed time which is called here time is the duration from one moment to another. Time is usually considered to be communal to any Euclidean space in Galilean relativity. Motion changes the existence reliability and stabilizes the mass point reducing the deviation in the existence probability distribution, which is variability. The variability comprises the elapse of time. A mass point with high inconstancy is apt to change and together with high insecurity, it becomes more changeable. For variability insecurity and inconstancy are in the same direction. Insecurity and inconstancy are proportional in variability. Variability is defined as the quotient of existence probability and persistence probability. As described later, the variability (Z) is a constant and $Z = \Phi/\Psi$.

2. Continuity of a mass point

The substantial stability consists of variability and continuity. Natural events carry on or change. Changes are essentially for stability which is continuity. When situations change, events vary for continuity. For continuity events with high insecurity have to have low inconstancy and events with high inconstancy have to have low insecurity. In the continuity of events insecurity and inconstancy are in opposite direction. Hence insecurity and inconstancy are inversely proportional in continuity. Persistence reliability and existence reliability are inversely proportional. Continuity is defined as the product of existence probability and persistence probability. As described later, the continuity (H) is a constant and $H = \Phi^*\Psi$.

3. Substantiality of a mass point

The actuality of events is usually perceived with scholar values. But the values should be considered as statistically the most possible values which are determined by the balance of variability and continuity. When an event is in changing situation, the substantiality has large variability and its continuity should be small. And when it is in the stabilized condition, its variability is small and its continuity is large. Variability and continuity are in opposite direction. They are inversely proportional in substantiality of events.
Usual human perception often makes the following statement.

"Fragile things are hard to exist, which can, however, exist in favorable situation."

This expression can be explained as follows. Events with low continuity easily change having high variability but in the situations with low variability, they have high continuity. Events with high continuity have low variability. Hence the substantiality is the product of variability and continuity. Substantiality is defined as the product of variability and continuity.

4. Uncertainty of a mass point

A quick change in certain energy loss requires tiny time (Δt) which can't be zero. Even a lasting event accompanies slight energy change (ΔE) which can't be zero. The product of ΔE and Δt is never zero, which has been known in quantum level as the uncertainty principle. This formula ($\Delta E^*\Delta t \geq h$) is demonstrated as follows.

The differentiation of momentum is force (dP/dt = F). $\Delta P/\Delta t$ is considered here the approximation of dP/dt, therefore,

$$\Delta t \fallingdotseq \Delta P/F$$

Equation (8) is,

$$P = E/[\tau(2\pi)^{1/2}]$$

$$\Delta P = \Delta E/[\tau(2\pi)^{1/2}]$$

Equation (3) is,

$$F = E/[\sigma(2\pi)^{1/2}]$$

therefore,

$$\Delta t = \Delta E/E/\omega \quad \omega = \tau/\sigma$$

$$\Delta E \Delta t = (\Delta E/E) * \Delta E/\omega$$

When E is very small and E can be considered close to ΔE, then $\Delta E/E \fallingdotseq 1$.
$\Delta E = nh\omega$, therefore,

$$(11) \quad \Delta E \Delta t = nh$$

Hence,

$$\Delta E \Delta t \geq h$$

But when E is large,

$$\Delta E/E \fallingdotseq 0$$

then,

$$\Delta E \Delta t \fallingdotseq 0$$

The proposition of uncertainty is applicable to very small mass points which may be the quantum level.

F. Actuality of two mass points

1. Linear and spatial occurrence

The occurrence of energy is uniform in a Euclidean space. Hence the occurrence on a line in the space is also uniform. In consequence existence probability of a mass point forms a normal distribution. Let P_C be the linear occurrence of a unit of energy and let E be the amount of energy of a mass point. Its existence probability is a normal distribution and the relationship of standard deviation (insecurity) and occurrence is shown with equation (5) which is $\sigma^2 \fallingdotseq P_C/E$. Hence the square root of the quotient of linear occurrence and energy is the standard deviation which is the insecurity of mass point. The occurrence of a mass point in a space is all the same at any point. Let ε be the standard deviation of existential probability spatial distribution of energy E. Then the occurrence of energy E in a unit volume is $E*\varepsilon^2$. And its relationship to the linear occurrence (P_C) is as follows, where Δs is the cross section in the vicinity of x-axis.

(12) $\qquad P_C*dx = \Delta s*E*\varepsilon^2*dx$

$\qquad\qquad E*\sigma^2 = \Delta s*E*\varepsilon^2$

$\qquad\qquad \sigma^2 = \Delta s*\varepsilon^2$

(13) $\qquad \Delta s = \sigma^2/\varepsilon^2$

The dimension of Δs is area and dimension of σ^2 is square of the distance, therefore, ε has null dimension.

2. Occurrence of another mass point

The occurrence of mass point (A) is all the same at any point in a space.

F. Actuality of two mass points

Hence linear occurrence (P_C) is uniform at any point over x-axis. But when there is another mass point at the origin, the appearance rate of mass point (A) on x-axis is not uniform over the axis. Mass point (A) stays at (x, 0). The appearance rate of origin point from mass point (A) is the configuration factor. The angle is regarded as proportional to the oscillation angular velocity (ω) and the solid angle is regarded now as a circular cone. The factor (Ro) is the ratio of base surface $\pi(a\omega x/2)^2$ and spherical surface ($4\pi x^2$).

$$Ro = \pi(a\omega x/2)^2/(4\pi x^2) \qquad a \text{ is proportional constant}$$

$$= (a\omega)^2/16$$

Ro is constant to x. The appearance of mass point (A) from the origin (Rox) is proportional to the quotient of cross section (Δs) and the spherical surface. Δs is replaced with equation (13).

$$Rox = b\Delta s/(4\pi x^2) \qquad b \text{ is proportional constant}$$

$$= (b\sigma^2)/(\varepsilon^2 * 4\pi x^2)$$

Hence occurrence (R_x) of mass point (A) at point (x, 0) is as follows.

$$R_x = Ro * Rox$$

$$= (a^2\omega^2/16) * (b\sigma^2)/(\varepsilon^2 * 4\pi x^2)$$

$$= (a^2\omega^2 * b\sigma^2)/(G_1 * x^2) \quad \text{where } G_1 = (8\varepsilon)^2 \pi/(a^2 b)$$

(14) $\qquad R_x = \tau^2/(G_1 * x^2)$

Occurrence of mass point (A) at point (x, 0) is proportional to τ^2 and inversely proportional to x^2.

3. Influence and rotation of another mass point

When a mass point (A) exists in a space, it can stay at any point in the space since occurrence is uniform. Now a mass point (A) stays at point $(x, 0)$ which is stable. If another mass point would exist in the space, it is not stable. They influence each other. The existence of mass point (A) at point $(x, 0)$ is no more stable. The influence of mass point (A) to the mass point at the origin is considered here. The more distance from origin it has, the more stable it is. The occurrence of mass point (A) at point $(x, 0)$ in relation to the mass point at origin is inversely proportional to the square of distance x as expressed with equation (14).

A stochastic variable (u) of which occurrence of the mass point (A) is constant at any value of (u) is postulated. When its value is 0, x has to be infinite. While value (x) increases from (x) to infinite, value (u) has to decrease from (u) to 0. Meantime integrated occurrence of both variables has to be the same. Occurrence rate is constant as long as (u) concern. Let q be the occurrence rate of (u). When $q*u*du$ is the increase, occurrence rate decreases with (x). R_x is the occurrence rate at x, then the decrease at x is R_x*dx.

(15) $\qquad R_x = \tau^2/(G_1*x^2)$

Both sides are integrated. (x) is from infinite to (x) and (u) is from 0 to (u).

$$\tau^2/(G_1*x) = qu^2$$

Hence the square of the variable (u) is inversely proportional to x.

(16) $\qquad u^2 = q_x/x \qquad\qquad q_x = \tau^2/(G_1*q)$

Then mass point (A) makes the normal distribution with variable (u). The mass point (A) is the most stable when x is infinite, which means when u is 0. The occurrence is constant at any value of (u), which means the mean of (u) value is 0. Existence reliability of mass point (A) is the maximum at $u = 0$. Hence insecurity probability distribution with stochastic variable (u) forms normal distribution and is as follows.

(17) $\quad \Phi = \exp[-u^2/(2\eta^2)]/[\eta(2\pi)^{1/2}]$

η is the standard deviation. Equation (16) is substituted to this equation. Then equation (18) is obtained.

(18) $\quad \Phi = \exp[-1/(2x\eta^2)]/[\eta(2\pi)^{1/2}]$

This is the existence probability distribution on the x-axis. Persistence distribution is as equation (7). It is a normal distribution centering 0.

(7) $\quad \Psi = \exp[-V^2/(2\tau^2)]/[\tau(2\pi)^{1/2}]$

The continuity is applied to these two equations and the equation below is obtained.

(19) $\quad d^2x/(dt)^2 = -(\tau^2/\eta^2)/x^2$

Equation (19) shows that a mass point at the point (x, 0) has the centripetal acceleration toward the origin as gravity which is inversely proportional to square of distance (x).

From equation (14), when $x = 1$,

$R_1 = \tau^2/G_1$

From equation (5),

$\eta^2 = R_1/E$

$\eta^2 = \tau^2/(G_1*E)$

The above equation is substituted by equation (19).

(20) $\quad -d^2x/dt^2 = (G_1*E)/(x^2)$

Hence a mass point at $(x, 0)$ has the acceleration toward origin which is proportional to energy and inversely proportional to square of distance as the universal gravity.

4. Rotation

Continuity makes centripetal force as equation (19) describes. The quotient of inconstancy and insecurity of equation (19) is angular velocity which means the mass point at (a) rotates around the rotation center.

Variability should be applied to equations (18) and (7). The resulting equation is as follows and describes the centrifugal acceleration.

$$(21) \qquad d^2x/dt^2 = (\tau^2/\eta^2)/x^2$$

The positive acceleration causes centrifugal force. Continuity provides the centripetal force of the same quantity as described in equation (19), therefore, the distance between the mass point and the rotation center does not change in rotation.

G. Basic propositions

1. The universal gravity

The existence of a unit of energy at the origin point has its existence probability distribution with insecurity (σ_1), which is $\exp[-X^2/(2\sigma_1^2)]/[\sigma_1(2\pi)^{1/2}]$ and causes the centripetal force in its vicinity. A mass point of energy K at the deviation X receives the centripetal force toward $X = 0$. The centripetal force ($-F$) of energy K is as follow,

(22) $\quad -F = K\exp[-X^2/(2\sigma_1^2)]/[\sigma_1(2\pi)^{1/2}]$

The energy incidence at $X = 0$ multiplies this centripetal force. When the energy E resides at $X = 0$, the existential probability distribution of energy K determines the existential incidence of energy E at $X = 0$, which is $E\exp[-(-X)^2/(2\sigma^2)]/[\sigma(2\pi)^{1/2}]$ and modifies the centripetal force of energy K at X.

The product of this energy incidence and equation (22) is the centripetal force ($-F$).

$$-F = E\exp[-(-X)^2/(2\sigma^2)]/[\sigma(2\pi)^{1/2}]$$

$$*K\exp[-X^2/(2\sigma_1^2)]/[\sigma_1(2\pi)^{1/2}]$$

$$= KE\exp[-(-X)^2/(2\sigma^2)-X^2/(2\sigma_1^2)]/(2\pi\sigma_1\sigma)$$

$$= KE\exp[-X^2(\sigma^2+\sigma_1^2)/(2\sigma^2\sigma_1^2)]/(2\pi\sigma_1\sigma)$$

When X is large enough, the approximation is done applying Taylor expansion.

(23) $\quad -F = KE[2\sigma^2\sigma_1^2/X^2(\sigma^2+\sigma_1^2)]/(2\pi\sigma_1\sigma)^2$

$$= KE[\sigma\sigma_1/(\sigma^2+\sigma_1^2)]/(\pi X^2)$$

Newton's equation of the universal gravity is obtained.

$$-F = KE\{\sigma\sigma_1/[\pi(\sigma^2+\sigma_1^2)X^2]\}$$

(24) $\quad -F = G*KE/X^2$

G is the gravitation constant which is $G = \{\sigma\sigma_1/[\pi(\sigma^2+\sigma_1^2)]\}$.

2. Coulomb force

The existence of a unit energy mound causes its existence probability distribution with fragility and centrifugal force in its vicinity. A mound of energy E, which should be considered as + Coulomb, at the deviation X makes the centrifugal force against X = 0. The centrifugal force (F, toward positive) of energy E is obtained applying equation (22). But when the energy at X = 0 is an energy mound (K), its existential incidence due to the existence probability distribution of energy E modifies the centrifugal force of energy E. Therefore, the centrifugal force of energy E at X is the product of the energy incidences of energy E and K. In this case, σ is communal and F is the centrifugal force in Coulomb force. When X is large enough, the approximation is done applying Taylor expansion as is done in equation (23). The equation of the Coulomb force is obtained as follows.

$$F = KE/(2\pi X^2)$$

When the energy at X = 0 is an energy deficit (−K), the above equation expresses centripetal force, which is,

(24)′ $\quad -F = KE/(2\pi X^2)$

The negative Coulomb draws the positive Coulomb. Hence Coulomb (q) of energy (E) is,

$$q = E/(2\pi)^{1/2}$$

3. Mass and energy

A mass point in standstill has the inconstancy distribution and maintains corresponding energy. A mass point of motion decreases its inconstancy and increases its energy (E) as the motion velocity increases. The velocity has the limitation which is the maximum velocity (C). At the speed limitation energy increases no more.

Both sides of equation (8) are multiplied by C. CP has the dimension of energy.

$$CP = CE_1/[\tau(2\pi)^{1/2}]$$

CP is the total energy (E) which is $C/[\tau(2\pi)^{1/2}]$ times of E_1.

(25) $E = CP$

Let E_0 be the static energy and E be the total energy of a moving mass point.

From equation (25) $E_0 = CP_0$. Equation (10) is changed as equation (26).

(26) $E = E_0/[1-(v/C)^2]^{1/2}$

The approximation of equation (26) is as follows.

$$E = E_0*[1+(v/C)^2/2]$$
$$= E_0+(P_0/C)v^2/2$$

The dynamic energy is $(P_0/C)v^2/2$ which should equal to the dynamic energy by Newton $(m_0v^2/2)$.

$$E = E_0+m_0v^2/2$$

Hence,

$$m_0 = P_0/C$$

Both sides of equation (10) are divided by C,

$$P/C = (P_0/C)/[1-(v/C)^2]^{1/2}$$

Then $m_0 = P_0/C$, m_0 is static mass and, therefore, $P/C = m$, m is the mass of moving mass point.

From equation (25) $E/C = P$, then $E/C^2 = m$. Both sides of equation (26) is divided by C^2,

$$E/C^2 = (E_0/C^2)/[1-(v/C)^2]^{1/2}$$

The total energy divided by the square of the maximum velocity (E/C^2) is the total mass, therefore, following equation (27) is obtained.

(27) $\qquad m = m_0/[1-(v/C)^2]^{1/2}$

m is mass, m_0 is static mass

4. Stabilizing events and revolution

(1) Substantiality

Events are incorporated and actualized. Events have change and changelessness. The substantiality of an event comprises variability and continuity. A fragile event has large variability and small continuity. And a strong event stays still. Its variability is small and its continuity is large. In the substantiality (Θ) of an event variability (Z) and continuity (H) are inversely proportional, which is expressed with the following equation where substantiality (Θ) is a characteristic of an event which is a constant.

(28) $\qquad Z*H = \Theta$

Logarithm of both sides are taken,

$$\mathrm{Log}(Z) + \log(H) = \log(\Theta)$$

And both sides are differentiated, $d\Theta/dt/\Theta$ is 0.

(29) $dZ/dt/Z + dH/dt/H = 0$

As long as time dependent conditions do not change, continuity (H) does not change. Hence $dH/dt/H$ is 0. In consequence $dZ/dt/Z$ is also 0. H and Z are constants.

(2) Variability

Events vary. Fragile events are hard to be conserved, and they easily change with an insecure situation. When the inconstancy is high, events easily vary and with a high insecurity they vary even more easily. For variability existence reliability and persistence reliability are in the same direction. They are proportional in variability. Hence the variability is the quotient of insecurity probability (Φ) and inconstancy probability (Ψ).

Variability (Z) is the proportional constant as shown below.

(30) $\Phi/\Psi = Z$

Logarithm of both sides are differentiated,

(31) $(d\Phi/dt)/\Phi - (d\Psi/dt)/\Psi = (dZ/dt)/Z$

The variability induces that the remainder of insecurity variation rate and inconstancy variation rate is variability variation rate. Equations (2) and (7) are substituted into the equation above, then the equation below is obtained.

(32) $(dX/dt)X/\sigma^2 - (dV/dt)V/\tau^2 = (dZ/dt)/Z$

X is a deviation from point (a) on x-axis and $V = dX/dt$. Z is a constant. Then the equation below is obtained.

(33) $d^2X/dt^2/\tau^2 - X/\sigma^2 = 0$

Let $\tau/\sigma = \omega$, then, the dimension of ω is T^{-1}, which is the dimension of angular velocity.

$$d^2X/dt^2 = \omega^2 X$$

The mass point at X is expected to be revolving and has positive acceleration which causes revolving centrifugal force. When the mass point is moving along x-axis, above equation (33) can be changed as follows.

$$[d/dt+(\tau/\sigma)][d/dt-(\tau/\sigma)]X = 0$$

Since $[dX/dt-(\tau/\sigma)X = 0]$ diverges to infinite, which is not realistic, therefore, the solution is as follows.

(34) $\quad dX/dt+(\tau/\sigma)X = 0$

The solution of this equation is following abducting function where X_0 is initial value of deviation X.

(35) $\quad X = X_0 * \exp[-(\tau/\sigma)t]$

The deviation reduces with the attenuation coefficient (τ/σ) and converges to 0, which means that the amplitude of wave function (38) converges to 0 as described next.

(3) Continuity

Continuity is the stabilized existence. Fragile events can continue with proper situation which is low insecurity. Sturdy solid events can continue even with high insecurity situation. For continuity high existential reliability which is low insecurity situation or high persistence reliability which is low inconstancy is required. They are in the opposite direction. The continuity (*H*) is the product of insecurity probability (Φ) and inconstancy probability (Ψ). The continuity (*H*) is the proportional constant. Continuity is waving. When a mass point (A) stays still at point (*a*) on x-axis, the continuity is,

(36) $\quad \Phi * \Psi = H$

Logarithm of the both sides are differentiated,

$$(d\Phi/dt)/\Phi + (d\Psi/dt)/\Psi = dH/dt/H$$

Equations (2) and (7) are substituted to the equation above and then the equation below is obtained.

$$(dX/dt)X/\sigma^2 + (dV/dt)V/\tau^2 = dH/dt/H$$

X is deviation from point (a) on x-axis. $V = dX/dt$ and H is a constant. Hence the equation above becomes the equation below.

(37) $\quad d^2X/dt^2/\tau^2 + X/\sigma^2 = 0$

The ratio of τ/σ is revolving angular velocity (ω). The equation above is an oscillating function. Then the equation below is obtained.

(38) $\quad d^2X/dt^2 = -\omega^2 X$

Equation (38) indicates a mass point has the material wave by De Broglie. This solution is a wave function which is material wave whose wave length (λ) should be $\lambda = h/P$. Light is a wave and also has characteristics of particles. Then particles should have characteristics of wave. The wavelength of an electron is less than one thousandth of that of visible light. Particles and mass points have latent waving. The material wave of a moving particle has progressive wave and is employed in Schrödinger's wave equation. These established facts support the proposition of substantiality.

(4) Revolution

When there would be a mass point at a point (b) in x-axis whose existence reliability would be lower than that of point (a), then the mass point make the motion to point (a) on x-axis. Point (b) is regarded as the deviation (X) of point (a). Centripetal force from point (a) affects to the mass point at point (b). The mass point makes motion according to the variability which is equation (34) and stabilizes according to the continuity which is equation (37). According to the substantiality, the sum of these two equations shows the motion.

Equation (34) shows the motion from point (b) to point (a) due to the variability. Equation (34) can be changed as follows.

(39) $\quad 2\omega dX/dt + 2\omega^2 X = 0$

The continuity contributes to the motion. Equation (37) is,

(37) $\quad d^2X/dt^2 + \omega^2 X = 0$

According to equation (29), the substantiality of motion is the sum of equations (39) and (37), which is the equation below.

(40) $\quad d^2X/dt^2 + 2\omega dX/dt + 3\omega^2 X = 0$

The solution of the equation above is as follows.

$$X = \exp(-\omega t)*[X_1*\cos(2^{1/2}\omega t) + iX_2*\sin(2^{1/2}\omega t)]$$

X has to be a real number, therefore, $X_2 = 0$.

When $t = 0$, X is initial deviation X_0, therefore, $X_1 = X_0$.

Then equation (41) is obtained which is an abducting wave function.

(41) $\quad X = X_0 \exp(-\omega t)\cos(2^{1/2}\omega t)$

Wave amplitude X reduces and becomes close to 0 and the mass point is stabilized. The stabilized mass point still has the angle velocity, which means it is revolving on point (a). As equation (33) indicates, the revolving mass point at the point (a) has positive acceleration which causes revolving centrifugal force. And as equation (37) indicates, the mass point has negative acceleration which causes revolving centripetal force. Both forces stabilize the mass point.

5. Uncertainty of events

(1) Uncertainty principle

Physical events always involve some dispersion and fluctuation. The probability of no dispersion or no fluctuation does not exist, which means events should involve some fluctuation. Insecurity with some dispersion and inconstancy with some fluctuation induces continuity and variability of events. Events continue and alter. Uncertainty principle is usually expressed $\Delta E * \Delta t \geq h$ or $\Delta P * \Delta x \geq h$. Topological distribution of energy is the force and the derivative from chronological differentiation of momentum is also force. $\Delta E/\Delta x \fallingdotseq F$ and $\Delta P/\Delta t \fallingdotseq F$, therefore, $\Delta E * \Delta t = \Delta P * \Delta x$. Both expressions are the same. Momentum is employed to elucidate this principle. Events have continuity and variability. Momentum also has continuity and variability. When momentum increases from P_0 to P, the increasing curve makes an abducting exponential curve, which is due to the variability. The variability explains the uncertainty principle by Heisenberg.

(2) Variability of momentum

Momentum (P) has also insecurity and inconstancy. The derivative from the differentiation of momentum is force, therefore, the argument of momentum inconstancy probability distribution is the force. Hence momentum and force compose continuity and variability of momentum. The variability of momentum is induced as follows.

The probability distribution of momentum dispersion is the normal distribution (Φ) with a standard deviation (σ_P) and with the expected momentum which is (P_0) in the center as is expressed with equation (1)′.

(1)′ $\quad \Phi = \exp[-(P-P_0)^2/(2\sigma_P{}^2)]/[\sigma_P(2\pi)^{1/2}]$

σ_P is the standard deviation of Φ

P_X shall be deviation from the expected momentum (P_0), then $P_X = P - P_0$.

$$(2)' \quad \Phi = \exp[-P_X^2/(2\sigma_P^2)]/[\sigma_P(2\pi)^{1/2}]$$

The fluctuation is the time dependent change rate of deviation (P_X) and has the dimension of force.

$$(6)' \quad F = dP_X/dt$$

The argument of a momentum inconstancy probability distribution is the differential of momentum which is force. Let (Ψ) be the inconstancy probability distribution.

$$(7)' \quad \Psi = \exp[-F^2/(2\tau_P^2)]/[\tau_P(2\pi)^{1/2}]$$

Since insecurity and inconstancy are proportional in variability. Variability (Z) is the quotient of insecurity probability (Φ) and inconstancy probability (Ψ) as equation (30) shows. Equations (2) and (7) are substituted into equation (31) and processed, and then equation (32)' is obtained.

$$(32)' \quad (dP_X/dt)P/\sigma_P^2 - (dF/dt)F/\tau_P^2 = (dZ/dt)/Z$$

P_X is deviation from point (P_0) on p-axis and $F = dP_X/dt$.

When Z is constant, the next equation is obtained.

$$d^2P_X/dt^2/\tau_P^2 - P_X/\sigma_P^2 = 0$$

$$[d/dt + (\tau_P/\sigma_P)][d/dt - (\tau_P/\sigma_P)]P_X = 0$$

Since $dP_X/dt - (\tau_P/\sigma_P)P_X = 0$ diverges, the solution is as follows where $\tau_P/\sigma_P = \omega_P$.

$$(42) \quad dP_X/dt + \omega_P P_X = 0$$

(3) Uncertainty and variability

Uncertainty principle is usually expressed as follows.

G. Basic propositions

$$\Delta x * \Delta P \geq n*h \quad n \geq 1$$

$\Delta x*\Delta P$ is differentiated.

$$d(\Delta x \Delta P)/dt = \Delta x * d\Delta P/dt + \Delta P * d\Delta x/dt$$

Variability is applied to Δx and ΔP. Δx can be regarded as the small displacement of x, and ΔP as the small displacement of P, and then $d\Delta x/dt = -\omega_1 \Delta x$ and $d\Delta P/dt = -\omega_2 \Delta P$ are obtained as equation (34) describes.

Hence,

$$(43) \quad d(\Delta x \Delta P)/dt = -\omega_1 \Delta x \Delta P - \omega_2 \Delta x \Delta P$$

$$= -(\omega_1+\omega_2)\, \Delta x \Delta P$$

Let ω_3 be the sum of $\omega_1+\omega_2$, then $\quad = -\omega_3 \Delta x \Delta P$

Δx and ΔP are positive quantities, therefore, $d(\Delta x \Delta P)/dt$ is a negative function and has the dimension of energy which is n time of energy quantum $h\omega$.

$$d(\Delta x \Delta P)/dt = -nh\omega$$

$$-\omega_3 \Delta x \Delta P = -nh\omega$$

ω and ω_3 are the same since they are the angular velocity of energy $d(\Delta x \Delta P)/dt$. And then,

$$\Delta x \Delta P = nh \quad n \geq 1$$

Hence,

$$\Delta x \Delta P \geq h$$

6. Characteristics of heat and light

Heat exists localized in a space consisted of energy. Heat is energy and is the most elemental existence. Heat is fairly localized with an insecurity (σ) in space and its existential reliability is considerably low but never flat in space, therefore, the existential insecurity (σ) should be large but limited. Heat slowly infiltrates with an inconstancy (τ) into any direction and easily changes. Hence its inconstancy (τ) is large. Heat exists with insecurity and inconstancy. Hence heat follows the variability and continuity. The angular velocity (ω = τ/σ) should be limited. Heat is energy. Hence energy is just a kind of wave. Applying the wave function of equation (38) describes the frequency of heat or energy which should have rather a low angular velocity. Angular velocity which is the intensity of energy goes up, then energy quantizes. And very high intensity energy particulates and forms particles of mass which are substances.

Light is quantized energy and exists in a space. Its energy wave has the wavelength of 4000~8000 Å, which is a material wave. Light quanta have its own insecurity and inconstancy. Light is a quantized existence and has variability and continuity. Light exists at a point in a space at an instance. Its existential reliability is very high and it is highly localized. The existential insecurity (σ) can be rather small. It disappears right away. Equation (9) elucidates that the inconstancy (τ) is zero or close to zero since light quanta have the maximum speed or speed close to it. Hence its angular velocity (ω = τ/σ) is close to zero. Therefore, equation (35) shows its variability does change little. It is close to a constant. Equation (38) describes its continuity which shows almost no wave. Light goes straight and is affected little by anything around. Variability equation (35) describes the variability reduces and is close to zero in an instance. Hence light has almost no variability and moves in a straight line in a space. Light quantum is integrated energy, of which material waves are confined in a substantial straight line. Light has substantiality whose components have extreme values. Light is the extreme substance.

7. Variability and Lorenz function

Lorenz function can be induced by the variability. Equation (33) is derived from the variability, which is shown below.

(33) $\quad d^2X/dt^2/\tau^2 - X/\sigma^2 = 0$

Both sides of equation (33) are integrated. C_o is a constant which is independent of time.

τ in the next equation above is the inconstancy of a moving space time frame and τ' in the equation below is the inconstancy of the fundamental space time frame.

$$\sigma^2 dX/dt - \tau^2 X^2/2 = C_o$$

$$\sigma^2 dX/dt - \tau'^2 X^2/2 = C_o'$$

This equation below is substracted from the equation above.

$$\tau'^2 - \tau^2 = 2(C_o - C_o')/X^2 \qquad \text{then}$$

$$\tau'^2 = \tau^2 - 2(C_o' - C_o)/X^2$$

The dimension of both τ'^2 and τ^2 are the square of velocity, therefore, $2(C_o' - C_o)/X^2$ is replaced with $f'v^2$. f' is a positive constant.

$$\tau'^2 = \tau^2 - f'v^2$$

As v increases, τ'^2 decreases. τ'^2 is 0 or more, therefore, when τ'^2 is 0, v has the maximum velocity which is C.

$$f' = \tau^2/C^2$$

$$\tau'/\tau = 1/(1 - v^2/C^2)^{-1/2}$$

Right side of the equation above shows Lorenz function.

The Theory of Stability

H. Special relativity and Lorenz function

1. Relativity of motion

The correcting term is right side of equation (10) which is $1/[1-(v/C)^2]^{1/2}$ and is Lorenz function. It is necessary to revise mass, energy, momentum, etc., and is designated here revising γ function. The value of γ function is 1 or more and is required even for coordinates conversion.

The motion of a mass point is considered applying a coordinates system based on an origin point and a set of three coordinates. A space which based on an origin (R) is named here the coordinates system (α). Another space (β) of which origin point (R′) is moving on the x-axis of space (α) with velocity (v). A mass point (A) staying still with inconstancy (τ) at a point in the space (β) which is moving in space (α) in direction of x-axis with inconstancy (τ') and velocity (v). Now inconstancy of a moving mass point in the space (α) is considered. The inconstancy probability distribution (Ψ) in space (β) does not always coincide to that in space (α) since the mass point is moving in space (α). The value of mean fluctuation velocity (V) is 0 in space (β) but in space (α) the mean fluctuation velocity in x-direction is affected by moving velocity (v) and is not 0. The convolution of the velocity (v) and a function (f) is regarded as the mean fluctuation velocity (fv). The square of inconstancy $(\tau')^2$ is the mean of variation $(V-fv)^2$ which is the difference of mean $(V)^2$ and $(fv)^2$ in the condition that approximate fv is close to τ. The mean of $(V)^2$ is τ^2.

$$(\tau')^2 \fallingdotseq \tau^2 - (fv)^2$$

When the coordinates system (β) stays still to the coordinates system (α), $v = 0$, then (τ') coincides to τ. As v increases, $(\tau')^2$ decreases. $(\tau')^2$ is 0 or more. Hence as equation (9), $f^2 = \tau^2/C^2$ is obtained.

H. Special relativity and Lorenz function

(44) $\qquad \tau' = \tau/[1-(v/C)^2]^{-1/2}$

Therefore, $\tau' = \tau/\gamma$. Ratio τ/τ' is equal to γ function. The inconstancy of mass point (A) observed from the coordinate system (α) is (τ') which is smaller than the inconstancy observed from the coordinates system (β) which is (τ). When $v = 0$, $\tau = \tau'$. The measurement scale is communal to each space. The dimension of inconstancy is velocity, therefore, the velocity observed on the coordinate system (α) becomes slower. "Velocity in moving space retards."

2. Relativity of length

The same discussion as the above is applicable to the insecurity (σ). A mass point (A) stays still at a point of a coordinates system (β) which is moving in the coordinates system (α) in direction x with velocity v. The insecurity probability distribution (Φ) on coordinate system (α) has to be considered. The insecurity in the coordinates system (α) is not necessarily the same as that in the coordinate system (β). The insecurity (σ') observed from coordinate system (α) has an additional condition that the mass point is moving fixed to the coordinate system (β) where the mean deviation is 0 and the insecurity is (σ). In coordinates system (α) the mass point is in motion with velocity (v), which affects its insecurity. The mean deviation is not 0 and makes small shift in moving direction. Let the mean deviation be the product of the velocity (v) and a function (g). The square of insecurity $[(\sigma')^2]$ in the coordinates system (α) is the mean variation which is mean of $(X-gv)^2$. This mean variation is the difference of square of insecurity (σ^2) in coordinates system (β) and the square of mean deviation which is $[(gv)^2]$. When gv is close to σ,

$$(\sigma')^2 \fallingdotseq \sigma^2 - (gv)^2$$

When coordinates system (β) stays still to coordinates system (α), (σ') equals to (σ). As (v) increases, (σ') decreases. But (σ') is 0 or more, therefore, (v) can have the maximum value which is C.

$$g^2 = \sigma^2/C^2$$

(g) is a constant and following equation (45) is obtained.

$$(45) \quad \sigma' = \sigma/[1-(v/C)^2]^{-1/2}$$

Hence $\sigma' = \sigma/\gamma$. Insecurity (σ) has the dimension of length, therefore, the length observed on coordinates system (α) is shorter than that in coordinates system (β). Moving length is shorter than that in the standstill. "Length in moving space shrinks."

3. Relativity of time

A space can be defined by one origin point and three basic directions, which is Euclidean space and pleural Euclidean spaces can be defined in a Euclidean space. One Euclidean space can move in relative to another origin point. A position is determined in various spaces by the sum of the position vector and vector of origin. There is no relativity of time in Galilean relativity. Time is communal throughout to all spaces.

The natural time, which is the present instance, is communal to all Euclidean spaces. The time elapsed from past to now can be considered as negative time and the time duration from now to future can be positive time. Then, time or time duration from one instance to another instance is measurable continuous physical variable. Past, present and future are considerable at the same time in elapse time or in continuous time. From one instance to another instance has a conceptual interval or a numerical time which is named here conceptual time. Natural time which is measured time has some instability to conceptual time. Time (t) at a point of coordinate system (α) of which origin is R is considered. The time difference between conceptual and natural time is not always 0. Mean of the difference (T) is 0. The difference probability distributes with a normal

H. Special relativity and Lorenz function

function (Y) centering 0, which is expressed by the following equation. The standard deviation is (v) of which dimension is time. This instability is applicable at any point in space (α).

$$Y = \exp[-T^2/(2v^2)]/(2\pi v^2)^{1/2}$$

v is the standard deviation of Y

The equation above is also applicable to any point in coordinates system (β) of which origin is R′ when it is stationary to space (α). But when origin R′ is in motion in relative to origin R, natural time in space (α) does not change but the instability of natural time in space (β) increases when observed in space (α). T will change. T′ is time in space (β) measured in space (α) and its standard deviation is v′. When observed from space (α), the mean of T′ cannot be 0. The mean depends on velocity. When coordinates system (β) is moving with velocity (v) in direction x-axis of space (α), the mean of T′ shall be hv, where h is a function. The variance of T and that of T′ are the same except the mean value of T′. Hence v^2 is the mean of $(T'-hv)^2$. The mean variance of $(T')^2$ is $(v')^2$, and the following equation is obtained.

$$v^2 \risingdotseq (v')^2 - (hv)^2$$

When $v = 0$, which means space (β) stays still in space (α) v = (v′). Instability coincides in both spaces. v^2 has to be 0 or more. Let C be the maximum value of v. Then h is constant and $h^2 = (v')^2/C^2$, therefore,

$$v^2 = (v')^2 * (1 - v^2/C^2)$$

(46) $$v' = v*[1-(v/C)^2]^{-1/2}$$

When origin R′ is moving in relative to origin R, $v' = v*\gamma$ which means "Time in moving space prolongs." Hence each space has to have its own time-axis.

4. Coordinate transformation

Coordinate transformation in x-axis from the coordinates system (α) to coordinates system (β) which is moving parallel to x-axis is considered.

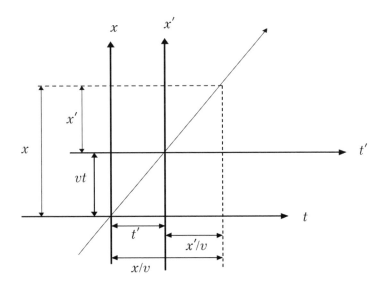

Let origin R of coordinates system (α) be on origin R′ of coordinates system (β) at one instance. After t seconds R′ moves on x-axis and come to vt. At this instance x' is on x. As the figure above describes, x' equals $(x-vt)$. But x' appears γ times shorter in coordinates system (α).

$$x'/\gamma = x-vt$$

therefore,

(47) $\quad x' = (x-vt)*\gamma$

Coordinates transformation to a coordinates system based on the moving origin requires transition of distance, time as well as γ factor. When the transformation

H. Special relativity and Lorenz function

is considered in time-axis, the sum of t' and x'/v equals to x/v. But time prolongs in moving coordinates system with factor γ, therefore,

$$(t'+x'/v)\gamma = x/v$$

$$vt' = x/\gamma - x'$$

$$vt' = x/\gamma - (x-vt)\gamma$$

$$vt' = [vt-x(1-1/\gamma^2)]\gamma$$

$$vt' = (vt-v^2x/C^2)\gamma$$

(48) $\qquad t' = (t-vx/C^2)\gamma$

The time transformation in coordinates system which based on the moving origin requires distance, velocity as well as γ function.

Epilogue

Man has had an ancient conception that God created the whole world and divine volition controls the whole or the majority of natural events. But the whole or the majority of nature is created by the contingency. The contingency brings forth various laws of nature which are always sought for to describe the nature. These two fundamental principles were realized in the processing a set of oncology data. Some fifty years ago author was trying to develop growth models processing clinical data and realized fluctuations of growth curves which were considered as measurement errors. They appeared waving and the growth curves appeared to contain wave functions, which easily explain the overshooting of the growth curves. Repeating long statistical processing the data, it is realized that the waving is explained when the stability of a mass is expressed with a normal function.

Author realized some thirty years ago, this concept explains the presence of the universal gravity law. Principle of insecurity deduces the universal gravity and principle of inconstancy deduces the special relativity. Variability and continuity of events are realized here, which substantialize the existence of events. Variability and continuity are formed from the above two principles.

Man can easily accept the following expression on the continuity of events.

"Fragile things are hard to exist, which can, however, exist in favorable situations."

Events of low continuity easily change with high variability but stay long with conditions of low variability. Events with low continuity will change into high continuity reducing variability with favorable conditions to exist easier.

Analyzing variability and continuity, the above stated two principles are developed. Summarizing this concept, these principles, variability and continuity are essential in the actual existence of events.

安定性理論

プロローグ

　人の考える存在すなわち概念上の存在は正確である。自然の存在は人間の思考とは独立の存在であり、自然の法則に従う、また偶然性が介在する。概念上の安定な存在とは質的な差異がある。各々の物体の存在やその動きは偶然であり、その偶然性のもたらす法則に基づいている。自然は偶然がもたらす安定化の産物であり、変動は安定化の原則に基づいている。自然のもたらす事象の本態を統計量として解析したのが安定性理論である。この安定性理論の本態は次の二つの原理から成る。

1. 散在性原理：事象は存在する。存在には統計的な散在性がある。
2. 揺動性原理：事象は持続する。持続には統計的な揺動性がある。

　事象は状況が変化すれば変動して安定化するため存続する。事象には変動性と存続性がある。安定化する変動には存在の確率と持続の確率は比例する。安定な存続には存在の確率と持続の確率は反比例している。事象の存在実態はこれら変動性と存続性の和集合である。事象は変動して安定化する。存続性の低い事象は存続し易い状況に向かって変動性を減少させ存続性を高める。

　上述の概念はこの安定性理論が諸般の基本的物理法則に矛盾していないことを示している。万有引力の法則は散在性原理に起因する。特殊相対性理論の成因も揺動性原理に在る。これら二つの原理から変動の法則と存続の法則は導かれ事象の変化と存在を説明する。事象は変動して安定化し存続する。変動性と存続性が反比例しているのが事象の存在実態である。この安定性理論を用いた偶然性に基づく理論は物理的基本法則

を説明することができる。この二つの原理が色々な事象に応用され、各種研究にも利用されることを期待する。変動と存続の概念は認識されるべきであり、多くの項目に応用され得ると考える。

目 次

　　　　プロローグ 61

A．存在の本質 67
1．空間、時間、エネルギー 67
⑴ 空間　　⑵ 時間　　⑶ エネルギー
2．存在の散在性と揺動性 69

B．散在性と引力 71
1．散在性原理 71
2．存在の散在性 71
⑴ ばらつき　　⑵ 一点上の存在
3．存在の確度分布 73
⑴ 存在確度分布　　⑵ 復元力
4．偏位と求心力 74
⑴ 存在精度と引力　　⑵ 生起率と運動

C．揺動性と相対性 77
1．揺動性原理 77
2．存在の揺動性 77
⑴ 持続性　　⑵ 揺らぎ
3．持続の確度分布 78
⑴ 持続確度分布　　⑵ モメンタム

4．速度と相対性 81

D．仮想的存在 83
　　1．仮想的エネルギー 83
　　2．仮想状態の力 83
　　3．実在と仮想状態 84

E．質点の実態 85
　　1．質点の変動性 85
　　2．質点の存続性 85
　　3．質点の実在性 86
　　4．質点の不確定性 86

F．質点間の実態 89
　　1．線上と空間の生起率 89
　　2．他の質点の生起率 90
　　3．他の質点への影響 91
　　4．旋回 93

G．基本法則 94
　　1．散在性と引力 94
　　2．クーロン力 95
　　3．質量とエネルギー 96

4．事象の安定化 97
　　⑴ 実在性　　⑵ 変動性　　⑶ 存続性　　⑷ 回転

5．事象の不確定性 102
　　⑴ 不確定性原理　　⑵ モメンタムの変動性
　　⑶ 不確定性と変動性

6．熱と光の特性 105

7．変動性とロレンツ関数 105

H．特殊相対論とロレンツ関数 107

1．運動の相対性 107

2．距離の相対性 108

3．時間の相対性 109

4．座標変換 111

　　エピローグ 113

A．存在の本質

1．空間、時間、エネルギー

(1) 空間

　空間が物理的に如何なる存在であるかは把握困難である。宇宙の大空間がどんな存在であるとしても、この近傍の空間は三次元空間であると考えられるゆえ、三次元空間、すなわちユークリット空間内での存在と考える。点の集合体である空間は基準点と基準となる３方向で定まる。これがこの基準点に準拠する空間でありユークリット空間である。一つのユークリット空間の中にもいくつかのユークリット空間を考え得る。一点の存在は一つのユークリット空間の中で表現することができる。

　一点の存在には位置と運動の概念があるが点の位置は相対的であり、運動とはその相対関係の変化である。線は点の連なりであり、一般的に点や線、形も点または点の集合体と考えユークリット空間内で表現できる。これらの点や点の集合体をここでは形態と称する。形態を構成する一点はユークリット空間内の一点に概念上では正確に位置をとり得る。これらの点の一つがユークリット空間内のある一点、例えば基準点に一致するなら概念上ではその点は完全な精度で、基準点に一致して存在するのである。また、その点が基準点以外の点に移るとその点と基準点の相対関係が変わる。この動きを運動と呼ぶ。これらの概念的な位置や運動は実在の位置や運動として把握されているが、ある程度不確実性を伴っているのが自然である。

(2) 時間

　時間は空間を構成する一要素と考える。時間は存在を構成する空間の一独立関数である。空間内で静止または運動している物体には独立の値

として時間がある、また空間内の他の要素、例えば基本方向の距離などは時間の従属関数となり得る。ゆえに空間と時間は存在の枠を構成する。ガリレオ相対論では時間はすべてのユークリット空間に共通で空間同士の関係では相対性はなく共通と考えた。すべての空間にはそれぞれの時間があり、互いに関連する相対性を有している。

　時間はエネルギーの構成要素であり、空間と時間は物理的事象の存在を表すのに必要である。時間も存在の一つの次元である。質点の存在は時間と空間の四次元からなる時空枠で表し得る。この時空枠をここでは時空と称する。また一つの時空内にいくつかの時空が存在し得る。各々の時空にはそれぞれの空間要素と時間がある。それぞれの時空が基本の時空に対して動きがない場合は、時間は共通で相対性はない。時間の相対性は変化する事象や運動する時空の間で起こる。

(3) エネルギー

　最も基本的な存在はエネルギーである。エネルギーには強度と量がある。エネルギーの存在にも位置と運動があり存在確率がある。すなわちエネルギーにも散在性と揺動性がある。エネルギーの存在は散在性原理や揺動性原理に従っている。エネルギーの揺動性はかなり高いと考えられる。エネルギーは変化し易いゆえ、変動性も大きいと考える。強度の高いエネルギーは光量子や質点となる。光量子や質点はエネルギー担体である。エネルギー担体にも位置や運動があり、存在確度分布がある。エネルギーの存在にも統計的な精度すなわち確度がある。その存在は散在性原理や揺動性原理に基づいている。強度の高いエネルギーは質量すなわち質点を構成する。エネルギーに関して質点の存在を把握する。

　ある一点が概念上の一点でなく物理的な存在の一点、すなわち質点であるなら、完全な精度で概念上の点に一致するとは言えない。確率的には少しずれている可能性もある。完全な一致ではなく確度を伴う。ずれ、すなわち偏位があり得る。散在性がある。質点の物理的存在位置は

A．存在の本質

存在確度分布の中心が安定で概念上の存在位置と一致する確率は高い。偏位は確度の低下を伴い、確度の高い中心への力が働く。この求心力が引力の起因となる。質点の運動も確度の変化を伴う。

　物体もまた点の集合である。それゆえその存在は時空枠、即ち時空にて表示し得る。時空での物体の存在が、人間の思考や概念的存在とは必ずしも一致しない。思考や概念は尺度でありスカラー量である。実在の物理量は統計量である。この意味に於いて不安定性を有したエネルギーの点や点の集合体として存在する。これが物体である。

　物体は位置と運動を有するが、種類と大きさにより運動速度の持続に難易の差がある。その度合いを示すものを質量と呼ぶ。運動速度の維持にも不確実性である揺動性がある。物体の存在を思考上ある一点で代表させる。形や大きさを無視して幾何学的な点とみなし、これに物理的な質量を与える。これが質点である。

2．存在の散在性と揺動性

　人間の考える位置すなわち点は概念上の存在である一点で、時空内の一点に正確に位置を取ることができ、また、その点は時空内を概念上自由に動き得る。完全な精度で、不確実性はなく、時空内の一点に存在し得る。また、運動にも完全な安定性がある。最も単純な存在である点状の物体で考えてみる。自然の存在と関連させた一点の場合、すなわち一質点の場合は人間の自由な思考通りには存在し得ない。また、動き得ない。人間の思考とは独立の存在であり、自然の法則に従う。人為的存在もあり得るが、自然の法則に従う存在には偶然性が介在する。物体上の一点は人間の思考に対して独立で物体固有の不安定性を有している。すなわち、物体上の一点は時空間内の一点上に存在するとしても、その存在には、概念上の一点の存在とは質的な差異があり、ずれや揺らぎが存在する。質点の存在は位置と運動を有するが、存在位置にはずれやばら

つきが起こり得る。これを散在性と呼ぶ。また、持続する運動にも不安定な揺らぎが起こり得る。これを揺動性と呼ぶ。

B．散在性と引力

1．散在性原理

　事象は存在する。存在には統計的な不安定性である散在性がある。存在には存在の確率分布がある。

　質点の存在がある点（a）上に期待されるとき、その偏位（X）は（$x-a$）。存在確率分布は正規分布であり、偏位（X）への存在確度（Φ）は、

$$\Phi = \exp[-X^2/(2\sigma^2)]/[\sigma(2\pi)^{1/2}]$$

期待値上への存在精度が完全なことはなく、散在性が常に伴っている。散在性は標準偏差値（σ）で表す。

2．存在の散在性

(1) ばらつき

　まずは比喩を用いて感覚的に存在の概念を考えてみる。ある平面上にゴルフボールを置くとする。その平面の座標を考え、そのボールの重心を平面の座標（0, 0）の位置に置いたと考えることができる。この存在は思考上点（0, 0）の位置に存在すると考えられるとしても、現実にボールを100回置いて、皆同じ位置に置けると考えるのは概念上可能であっても現実にはない。現実は必ず誤差を伴う。今、大学生がこの行為を行ったとしたら、ボールの位置の最尤値は点（0, 0）であるとしても、ある程度のばらつきを生じる。同じ行為を小学生が行えば、ボールの位置の最尤値はやはり点（0, 0）であろうが、もっと散在している。ばらつきは大きくなるであろう。ゴルフボールの重心の存在が点（0, 0）の位置であっても質的に違う存在である。いずれもボールの重心の位置が

点 (0, 0) で代表される存在である。中心は点 (0, 0) に存在しても、ある標準偏差値をもって、その周りに存在確度を分散している。前者のほうがより精確で小さな近傍内に集在している存在であり、分散値が小さい。重心点の最尤値 (0, 0) の近傍内への出現頻度が高いので存在確度が高く高精度な存在と表現する。後者の方が誤差の大きな精度の低い散在性の高い存在である。存在の精確性に差異があり、後者の標準偏差は大きくばらついているので、上記重心点は近傍内への出現頻度が比較的低い。

(2) 一点上の存在

　逆の方向から考えて、今ゴルフボールの重心が点 (0, 0) の上に在るとする。この重心がゴルフボールの存在を代表できると考える。ゴルフボールの重心を質点と考えて、この質点が空間内の一点 (0, 0) に存在するとする。この重心点が点 (0, 0) の上に存在していると考えるのは概念であって、現実に点 (0, 0) 上に在る確率はかなり高いが散在性もあり、ずれの可能性を否定できない。その存在位置に不確実性がある。誤差すなわち不確実性があり存在確度を伴った状態である。存在状態に精度を伴う。その質点の散在の確度分布では存在が期待される点 (0, 0) 上での確度が極大で、その点を中心に周りにより低い確度のばらつきがあることである。これを存在の散在性と呼ぶ。その精度に高い状態や低い状態があり得る。

　このことは粒子の位置に関しては理解し易い。電子等の粒子のある一瞬に於ける存在位置は算出できる。しかし、すべての粒子がその算出位置に存在しているとは言えない。その計算位置には大なり小なりの分散を伴う。それが散在性である。算出位置は最尤値であり粒子はその周りに散在するので統計的存在位置である。

3．存在の確度分布

(1) 存在確度分布

　存在の確度分布とは本来位置又はその偏位への存在の可能性で、偏位を確率変数とする確度分布である。本来の存在位置の確度が最尤値であり、これを中心に正規分布を成す。その最尤値に於ける存在の確度が存在確度であり存在の精確さである。これが高いほど安定な存在と言える。この確度分布の標準偏差値を散在性と呼び、これが小さいほど最尤値に於ける存在の精度が高い状態で、質点はより安定に存在する。この標準偏差値の逆数を精度と定義する。煩雑さを避けるためにその質点（例えばゴルフボールの重心点）が存在する点の x 軸方向のみの状態確率について考える。この質点の生起率は x 軸上どこでも均一であるので、その確度分布（Φ）は、ある標準偏差値をもって、期待値（a）を中心に正規分布をしている。

$$(1) \quad \Phi = \exp[-(x-a)^2/(2\sigma^2)]/[\sigma(2\pi)^{1/2}]$$

σ はΦの標準偏差値　$\sigma>0$

X を点 x とその質点の存在期待値（a）からの偏位とすると、$X = x-a$ だから、

$$(2) \quad \Phi = \exp[-X^2/(2\sigma^2)]/[\sigma(2\pi)^{1/2}]$$

存在位置すなわち期待値上での存在の確度は、この散在確度分布の $X = 0$ の時で、$1/[\sigma(2\pi)^{1/2}]$ である。この値が高いほど少しのずれで確度は急に低下するので、ずれの存在しない期待値（a）上への存在する精確な状態と言える。この標準偏差値（σ）が大きいほど存在の確度は下がるゆえ、（σ）は散在性を表し、その逆数（$1/\sigma$）を存在の精度と呼ぶ。

安定性理論

(2) 復元力

　質点の存在期待位置が点 (0, 0) であるとして、その標準偏差が比較的小さく、すなわち高精度で期待値上での存在確率が高い確実な存在や、精度が悪く、その存在確度の低い存在がある。前者は少しのずれで存在確度が大きく低下する。元の位置の存在確度が高いので、戻る確率が高い。偏位が起こり難い。すなわち期待値に向かう力がある。強い復元力が働く。後者は少しのずれでは存在確度の減少が比較的少ない。すなわち精度が低く、その期待値に向かう力、復元力は比較的弱いと考えられる。確実な存在状況とは復元力が強く物が常に所定の位置に確実に存在している状態である。不確実な存在状況とは復元力が弱く物がばらついてしまう状況で、物が本来の位置にある確率が比較的低い状態である。この場合は精度の低く散在性の高い状態である。期待値に向かう復元力を求心力と定義する。

4. 偏位と求心力

(1) 存在精度と引力

　静止状態では周囲に求心力が均等に存在するから静止を保っている。質点に小さなずれが生じたとする場合、存在頻度即ち存在確度は下がる。中心の方が高い存在確度なので復元する力が働く。存在確度の高い存在では求心力が大きく、存在位置がずれる可能性が少ない。存在確度の低い存在では求心力が小さいから位置がずれる可能性がある。存在確度の高い存在には強い求心力が伴っていて、小さなずれの力では動かない。存在期待値上の求心力が最も強く辺縁に行くほど存在確度の低下に伴い弱くなる。求心力は存在確度の、即ち偏位 X の関数であるが、存在期待値上では精度 ($1/\sigma$) に比例するとする。存在期待値上の求心力は中心力と定義するが、式 (2) から中心力 (F) は散在確度 $[1/\sigma(2\pi)^{1/2}]$ に比例する値とするゆえ、

(3) $\quad F = E/[\sigma(2\pi)^{1/2}]$

Eは比例定数でエネルギーのディメンジョンを持つゆえ、その質点のエネルギーと考えられる。その求心力は偏位Xの関数で引力の原因となる。偏位XにエネルギーKがあれば、エネルギーEの偏位Xに於ける存在確度分布によりエネルギーKに求心力をかける。同時にエネルギーKはX＝0の点にKの存在確度によるKに向かう求心力を起こす。この両者の積が引力となる。

⑵ 生起率と運動

質点のエネルギー量をEとし、単位エネルギー当たりの生起率をP_Cとすると、標準偏差値（σ）と生起率（P_C）の関係は、

(4) $\quad \sigma^2 = P_C(1-P_C)/E$

であり、P_Cは1に比して充分小さい値であるから、

(5) $\quad \sigma^2 \fallingdotseq P_C/E$

生起率を総エネルギーで除したものの平方根が標準偏差値となる。よって単位エネルギーの存在確度分布に於ける平均分散はおよそ生起率に等しいゆえ、生起率の均一な空間内の一点に存在する質点は、より高い存在確度の位置はないので安定であり、他の位置に移動を起こすことはない。たとえ起こしたとしても均一な空間であり、その散在性には変化がなく同じ中心力が作用する。質点に作用する求心力はどの方向からも均等であり、その位置に安定に存在している。この存在点上での散在確度が極大である。その存在確度は100％ではなく位置移動を起こす可能性も共存している。空間内の生起率が不均一で、質点の存在する位置（b）より低い生起率、即ち、より高い存在確度の位置（a）が発生するとすると、その位置はより高い存在頻度の位置なので質点は移動を起こす。

もとの位置 (b) はこの位置 (a) の散在確度分布に基づく偏位 X となり求心力により移動して安定化する。安定後の質点は存在確度が上がるので安定性はより高くなる。

C．揺動性と相対性

1．揺動性原理

　事象は持続する。持続には統計的な不安定性である揺動性がある。持続には揺動の確率分布がある。

　事象は運動を持続する。その持続は統計量であり統計学的な不安定性が内在する。それが揺動性である。持続確率分布は正規分布であり、揺動（V）での揺動確度（Ψ）は、

$$\Psi = \exp[-V^2/(2\tau^2)]/[\tau(2\pi)^{1/2}]$$

であり、期待値上への存在強度は完全なことはなく揺動性が常に伴っている。揺動性は標準偏差値（τ）で表す。

2．存在の揺動性

(1) 持続性

　同じ精度の存在としても変化し易い存在と持続する存在とがある。揺らぎを感覚的に考えてみる。ゴルフボールの代わりにピンポンボールを位置（0,0）に置く行為を行ったとする。大学生が上手に位置（0,0）に100回置いて、ゴルフボールの時と同じ高精度な存在を得られたとしても、ゴルフボールの場合より揺らぎ易い。置いたときの位置移動が起こり易い。すなわち定着性が悪い。ゴルフボールを置いた時の方が揺らぎが少なく静止し易い。静止の弾みが大きく、この位置での揺らぎが0である確率が高い。同じ精度の位置（0,0）上への存在であっても存在の定着性に差異がある。今ゴルフボールの重心が点（0,0）上に在り静止している存在であるとすると、その実態は揺らぎの速度が0、すなわち

固定である確度が最も高い存在である。固定の確度の高いほど静止の弾みが大きい存在である。次にゴルフボールの代わりにピンポンボールがあり、その重心が点 (0, 0) の上に在るとする。ゴルフボールに比して揺らぎの可能性は高く、揺らぎのない固定状態であるとしても静止の弾みは比較的小さく、存在位置への定着が弱い。ゴルフボールの方が落ち着きがあり、揺らぎの可能性の少ない存在と言える。

(2) 揺らぎ

存在には期待している点上にあるかどうかという位置的な不安定性すなわち散在性だけでなく、その点での揺らぎを起こす可能性の質的な不安定性も内在する。これを存在の揺動性と呼ぶ。揺らぎは持続すなわち揺動が0の固定速度を中心に正規分布を成す。その標準偏差値で揺動性を表す。その標準偏差値の逆数を強度と定義する。

強度が低く揺らぎ易い存在や、強度が高く揺らぎ難い存在がある。いわゆる重いボールは強度が高い存在で、静止の弾みが比較的大きく揺動を起こし難く持続する。ピンポンボールは静止の弾みが小さく揺動を起こし易い存在で、揺動が0の固定確度が低くなる。この相違は存在の強度の違いと表現することができる。強度が高いとは質点と位置の結びつき（静止の弾み）が強い状態と言える。点の集合としての形態の変化についても同様である。一つの形態の代表的な複数の点の相互関係が変化し難い状態が強度の高い存在である。一点上の存在を考えると、ずれの可能性だけでなく、揺らぎの可能性も在る。

一般的な存在には精度と強度の二つの安定要素がある。

3．持続の確度分布

(1) 持続確度分布

質点がある点上に確実に存在していることは、質点が一点上に持続し

て存在していることであり、静止（速度は0）していることである。この速度0にも不安定性があり、質点は揺らぐ可能性がある。質点が静止していることは、持続の確度分布で揺らぎなし（揺動速度0）の確度が極大で、その近傍の揺動はより低い確度で分布していることである。持続の確度分布の標準偏差値を揺動性と呼び揺らぎの可能性の大きさを表す。持続の確度分布とはその事象の揺らぎの可能性で、存在の確率変数（位置）の時間変化率を確率変数とする。この変化率をここでは揺動または揺動速度と呼ぶ。持続は揺動0が最尤値であり、その確度が持続確度である。その周りに正規分布を成す。その標準偏差値を揺動性と呼び、それが小さいほど静止確度が高く質点は弾みが大きく堅固な存在となる。この揺動の最尤値上での確度が存在の堅固さであるから、この標準偏差値の逆数を強度と定義する。揺動は偏位（距離）の時間変化率であり速度の次元を持つ。

　x軸上の点（a）の位置に存在する質点の偏位をXとすると揺動（V）は次の式で表される。

$$(6) \quad V = dX/dt$$

この質点の揺動は如何なる方向にも速度0を中心とした正規分布を成す。この確度分布（Ψ）が持続の確度分布である。この確度分布のばらつきの幅（標準偏差τ）が揺動性であり、物体の強度（持続の弾み）に反比例する。x軸上で考えると、$x = a$に静止している質点の揺動はある標準偏差値をもって揺動0を中心に正規分布を成すので、

$$(7) \quad \Psi = \exp[-V^2/(2\tau^2)]/[\tau(2\pi)^{1/2}]$$

$V = 0$のときの確度は$1/[\tau(2\pi)^{1/2}]$で、これが高いほど少しの揺らぎでその確度は急速に低下する。

(2) モメンタム

　ある質点が一つのユークリット空間内で、ある点上に安定して静止している場合は速度0である。質点がある点上に確実に存在していることは質点が一点上に持続して存在していることであり、静止（速度0）していることである。質点が静止していることは持続確度分布で揺らぎなし（揺動速度0）の確度が最大で、その近傍の揺動はより低い確度で分布していることである。揺らぎ難い存在とは静止のはずみが大きく堅固な状態の存在である。τはΨの標準偏差で揺動性を表すが、これが小さいほど$V=0$の確度が高くなり強度が増す。この標準偏差値の逆数（$1/\tau$）は堅固さに比例するゆえ強度と呼ぶ。

　エネルギー（E_1）と質点の持続確度の積、すなわち、点 a に静止している質点のエネルギー（E_1）と式（7）の$V=0$のときの確度$1/[\tau(2\pi)^{1/2}]$との積（P）はモメンタムの次元になる。

$$(8) \quad P = E_1/[\tau(2\pi)^{1/2}]$$

よってモメンタム（P）は強度に比例する、比例定数E_1はエネルギーの次元である。強度が高いほど静止モメンタムも大きく安定である。

　モメンタムと角速度の積は力であるから、中心力（F）と静止モメンタム（P）の比（F/P）も角速度（ω）である。FとPの比は式（3）と（8）から次式となる。

$$F/P = [E/(2\pi\sigma^2)^{1/2}]/[E_1/(2\pi\tau^2)^{1/2}]$$

$$= (\tau/\sigma)/(E_1/E)$$

（F/P）も（τ/σ）も（ω）であるゆえ、$E_1/E=1$となる。Eは散在性にも揺動性にも共通であり、静止モメンタムや中心力の大きさに係わる状態を規定する。また、その質点が代表する物体内のエネルギーである。

4．速度と相対性

　空間内の一点に存在する質点は、その位置より存在確度の高い位置があれば、その位置に移動する運動を生じる。運動が質点の揺動性に関係がないと考えるなら、運動中の質点の揺動（速度の次元）の平均は0で揺動の確度分布も変わりはない。しかし、運動している質点の運動方向の揺動性は変化する。速度（v）で運動している質点の持続確度分布は運動速度（v）の影響を受け、揺動の平均は0ではなく速度の方向にシフトする。シフトの平均を速度（v）にある関数（f）を乗じたものと置く。この場合の揺動性を示す標準偏差値を（τ'）とし、その二乗である平均分散（τ'）^{2}は持続確度分布の分散（$V-fv$）^{2}の平均である。fvがτに近い範囲では、この平均分散は式（7）に於ける静止時の平均分散（τ^2）から運動中のシフトの平均（fv）の二乗を減じたものとなるから、

$$\tau'^2 \fallingdotseq \tau^2 - (fv)^2$$

静止している時は、$v=0$であり、τ'^2とτ^2は一致する。vが大きくなるにしたがってτ'^2の値は小さくなるがτ'^2の値は0または正数でなければならないから、$(fv)^2$の最大値はτ^2である。ゆえにvの最大値をCとすると、

$$\tau^2 = (fC)^2$$

$$f^2 = \tau^2/C^2$$

fは定数となるので次式を得る。

(9) 　　　$(\tau')^2 = \tau^2[1-(v/C)^2]$

質点の運動速度には限界があり、この限界の速度は光速またはそれに近い速度と考えられる。この限界速度がアインシュタインの相対性理論の

光速だと考えられている。

　また、運動している質点のモメンタムを求めるには、式（8）より質点の静止モメンタム（P_0）は $E/[\tau(2\pi)]^{1/2}$ であり、速度 v の時のモメンタム（P）は $E/[\tau'(2\pi)]^{1/2}$ であるから、次式（10）を得る。

(10)　　　$P = P_0/[1-(v/C)^2]^{1/2}$

速度が上昇するにしたがってモメンタムは無限に大きくなり得るが、これは速度が限界に近づくからである。

D. 仮想的存在

1. 仮想的エネルギー

　式（2）に於ける標準偏差値が散在性である。この標準偏差値が増加するにつれて質点の存在確率は減少し、標準偏差値が無限大になれば質点の実質は消失する。式（3）が示す如く求心力も消失する。これが仮想状態への移行である。質点のエネルギーを示す式（3）に於けるEの存在確度分布は平坦となる。また、負の散在性も考え得る。散在性が負の仮想的存在も考え得るのである。そこではエネルギーEが仮想的存在を代表し、式（3）が示す如く遠心力が働く。エネルギー分布には山や谷ができることが考えられる。散在性は負の値であるから式（3）からわかるように、エネルギーの山（E）には遠心力が働き崩れ易く、谷（–E）には求心力が働き埋まり易い。

2. 仮想状態の力

　仮想質点に式（2）を適用する。存在確度は負になるが仮想質点は式（2）に当てはまる。そのXは仮想質点すなわちエネルギーの山または谷の中心からの偏位である。

$$(2) \quad \Phi = \exp[-X^2/(2\sigma^2)]/[\sigma(2\pi)^{1/2}]$$

エネルギーの山が為す遠心力は標準偏差の絶対値が小さいほど強力である。エネルギーの谷はエネルギーを引き寄せる。だから谷は平らになる傾向がある。山は崩れる傾向がある。式（3）をエネルギーの山に当てはめればFは負になり、遠心力を表す。山の中心点（X = 0）の虚弱さは$1/[\sigma(2\pi)^{1/2}]$である。この絶対値が大きいほど虚弱さは大きい、すな

わち平らになる力が強くなる。標準偏差の絶対値が減少するほど、虚弱さは増加する。だから標準偏差の逆数は中心部の負の強度、平らになる力を表す。遠心力はXの関数で、その中心はその強度に比例する。式(2)から負の中心力は $\{1/[\sigma(2\pi)^{1/2}]\}$ に比例する。

$$(3) \quad F = E/\{1/[\sigma(2\pi)^{1/2}]\}$$

比例定数Eはエネルギーである。Eが単位エネルギーの場合は偏位Xに於けるエネルギー分布率を表す。−Fは中心部の平らになる力を表す。

3．実在と仮想状態

　物体はエネルギー量子から構成されている実質的な存在である。例として中性子は実質的な存在である。中性子は陽子と電子になり得る。それらはエネルギー量子から成り立っている。中性子は陽子と電子になり得るが、双方とも完全にバランスの取れたエネルギー量子から成り立った存在に分裂するとは言えない。仮想的存在である余分なエネルギー量子が陽子に残り、エネルギー量子の不足が電子側に起こる。これも仮想的存在である。それゆえクーロン力が発生し、電子はある角速度をもって陽子の周りを回転し存続するのである。高振動数のエネルギーは量子又は粒子となる。粒子にはそれ特有のエネルギーレベルがあるのでエネルギーの過不足も起こり得る。この過不足状態がエネルギーの仮想状態である。

E．質点の実態

1．質点の変動性

　形態は思考上ユークリット空間内を自由に動き得る概念上の存在であるが、物体を構成する形態がユークリット空間を自由に動き得るわけではない。存在確度の相違があるからである。運動とは基準点と形態上の各点との相対関係の時間的な変化であり、存在確度の変化を伴って変位を起こす事象の変動である。運動には時間を伴う。運動を考えるにあたっては時間の概念が必要となる。ある時刻から他の時刻への経過が時間であり、一般的に（ガリレオ相対論では）如何なるユークリット空間でも共通であると考えられている。変動は存在確度分布の偏位による運動であり、その結果として存在確度の変化も生じて存続する。散在性が高い事象は変化し易いが揺動性が高ければ、なお変動し易い。これが変動性である。変動性は存在確度と持続確度の比である、と定義する。変動性（Z）は $Z = \Phi/\Psi$ であり、（Z）が定数であることは後述する。

2．質点の存続性

　安定な存在が実在であり変動性と存続性から成る。変動の本質は安定化である。散在性が高い事象でも、揺動性が低ければ存続し易いのが存続性である。事象は変動するか存続する。事象は状況が変われば変動して存続する。事象が存続するには散在性が高い時は揺動性が低くなければならない。揺動性が高い時は散在性が低くなければならない。存続性に於いては散在性と揺動性は逆方向である。よって、存続性では散在性と揺動性は反比例している。存続性は存在確度と持続確度の積である、と定義する。要約すると、事象はその散在性と揺動性に比例して

変動し、散在性と揺動性が反比例して安定に存続する。存続性 (H) は $H = \Phi*\Psi$ であり、(H) が定数であることは後述する。

3．質点の実在性

事象の実態は一般にスカラー値で把握される。しかし、それらの値は変動性と存続性のバランスで定まる統計的確率が最も高い値と把握するべきである。事象の実質は変動性と存続性とから成るが、変化しつつある事象の実在性は大きな変動性と小さな存続性から成る。また、安定した事象の実在性は大きな存続性と小さな変動性から成る。変動性と存続性は逆方向である。変動性が高い時は存続性が低く、変動性が低い時は存続性が高いのが事象の実在性である。変動性と存続性は反比例していて変動性と存続性の積が事象の実在性である。
物の安定性について次のことが言える。

　　　「壊れ易い物は存在し難いが、適切な条件下では存在する。」

この意味は、存続性の低い事象は変動するが変動性の低い状況を得れば存続するということである。存在状況が変化すると、事象は変化して安定化する。存続し易い安定な存在状態に変動する。この経験則は物理現象のみでなく多くの自然現象に適応できるであろう。また、人間の社会現象や精神現象等にも適応できるであろうが、ここでは質点の存在と運動に関して考える。

4．質点の不確定性

頑丈な物体もいずれ変動する。それには少量のエネルギー消失（ΔE）を伴う。少量のエネルギー消失での大きな変化でも短い時間が必要で、

変化時間（Δt）は 0 ではない。変化にはエネルギー（ΔE）と時間（Δt）を必要とする。いずれも 0 ではない。このことは量子レベルでの不確定性原理として知られている。$\Delta E * \Delta t \geqq h$ は次のように求められる。

モメンタムの微分は力である（$dP/dt = F$）。ここでは $\Delta P/\Delta t$ を dP/dt の近似値とする。よって、

$$\Delta t \fallingdotseq \Delta P/F$$

式（8）は、

$$P = E/[\tau(2\pi)^{1/2}]$$

上式から、

$$\Delta P = \Delta E/[\tau(2\pi)^{1/2}]$$

式（3）は、

$$F = E/[\sigma(2\pi)^{1/2}]$$

よって、

$$\Delta t = \Delta E/E/\omega \qquad \omega = \tau/\sigma$$

$$\Delta E \Delta t = (\Delta E/E) * \Delta E/\omega$$

E が大変小さくて ΔE に近い場合には、$\Delta E/E \fallingdotseq 1$。
$\Delta E = nh\omega$ であるから、

$$(11) \qquad \Delta E \Delta t = nh$$

よって、

$$\Delta E \Delta t \geqq h$$

安定性理論

しかし E が大きい場合は、

$$\Delta E/E \fallingdotseq 0$$

よって、

$$\Delta E \Delta t \fallingdotseq 0$$

この法則は質量が大変小さい場合に、おそらく量子レベルに適応できる。

F．質点間の実態

1．線上と空間の生起率

　質点が一空間内の如何なる点上でも存在（生起）する確率は均一である。よって、空間内の一直線上での質点の生起率も均一となる。x軸上の生起率P_Cはどの位置でも一定であるから、x軸上の一質点の偏差の確度分布は正規分布を成し、式（1）となる。単位質量当たりの線生起率をP_Cとし、ある質点のエネルギーをEとする。この質点の存在の確度分布は正規分布を成し、その標準偏差値（散在性）と生起率との関係は式（5）から$\sigma^2 ≒ P_C/E$、よって線生起率と質量の比の平方根が標準偏差値であり、質点の散在性である。質点が空間内の如何なる点にも生起率は均一であるから、単位エネルギーが空間内に生ずる平均分散をε^2と置くと、エネルギーEの単位容積内生起率は$E*\varepsilon^2$である。また、線生起率（P_C）との関係はx軸近傍の断面積をΔsとすると、

(12)　　　$P_C*dx = \Delta s*E*\varepsilon^2*dx$

　　　　　$E*\sigma^2 = \Delta s*E*\varepsilon^2$

　　　　　$\sigma^2 = \Delta s*\varepsilon^2$

Δsは面積の次元を持ちσ^2は距離の二乗の次元を持つからεは次元を持たない。線生起率と空間生起率の比は、

(13)　　　$\Delta s = \sigma^2/\varepsilon^2$

安定性理論

２．他の質点の生起率

質点（A）は空間内どこでも生起率は一定であるから、線生起率で考えるなら x 軸上どこでも一定の線生起率（P_c）である。だがもう一つの質点が原点 $(0, 0)$ に存在していて、これに対して x 軸上の点 $(x, 0)$ に質点（A）が現れる割合は一定ではない。その生起率は距離とともに減少してゆく。まず、x 軸上点 $(x, 0)$ に存在する質点（A）から見て原点が見える割合（Ro）は質点（A）から原点への視角の成す円錐の底面と原点を中心とする半径 x の球面との比である。視角は角速度である τ と σ の比すなわち振動角速度（ω）であり、これの成す円錐底面と球面積の比は、

$$\text{Ro} = \pi(a\omega x/2)^2/(4\pi x^2) \qquad a\text{は比例定数}$$

$$= (a\omega)^2/16$$

となり、x に関係なく一定となる。また、原点から半径 x の位置に質点（A）が生ずる割合は点 $(x, 0)$ 近傍の断面積 Δs と半径 x の球体表面積 $4\pi x^2$ との比となる。よって、原点から見た点 $(x, 0)$ 近傍内への出現割合を Rox とすると、

$$\text{Ro}x = b\Delta s/(4\pi x^2) \qquad b\text{は比例定数}$$

$$= (b\sigma^2)/(\varepsilon^2 * 4\pi x^2)$$

ゆえに、点 $(x, 0)$ に存在する質点（A）の原点に対する生起率（R_x）は、

$$R_x = \text{Ro} * \text{Ro}x$$

$$= (a^2\omega^2/16) * (b\sigma^2)/(\varepsilon^2 * 4\pi x^2)$$

$$= (a^2\omega^2 * b\sigma^2)/(G_1 * x^2) \quad \text{where } G_1 = (8\varepsilon)^2\pi/(a^2 b)$$

(14)　　　$R_x = \tau^2/(G_1 * x^2)$

点 $(x, 0)$ に存在する質点（A）の原点に対する生起率は τ^2 に比例し、x^2 に反比例する。

3. 他の質点への影響

　質点（A）がユークリット空間内に単独で存在する場合、その生起率はどこでも一定であり、どこにでも安定して存在し得る。x 軸上の点 $(x, 0)$ に存在する質点（A）はその場所に安定に存在し得るが、空間内にもう一つの点が存在する場合はお互いの影響があり質点（A）の存在は安定でない。点 $(x, 0)$ に存在する質点（A）が原点に存在する質点（O）に及ぼす影響について考えると、お互いの距離が離れているほど影響は少ない。質点（A）が最も安定に存在し得る距離は無限遠で、一点の存在に同じである。また、原点から見て質点（A）が x 軸上の点 $(x, 0)$ に存在する生起率は式（14）の如く x の二乗に反比例する。

　よって二乗が x に反比例する確率変数（u）を考えるなら、質点（A）の生起率は変数（u）の値に関係なく下記のごとく一定となる。x が x から無限大まで変化する間に u は u から 0 まで減少する。この間の生起率の積分値は一致しなければならない。x の変化量と u の変化量の関係を生起率から考えると x の生起率は x の増加とともに減少していくが、その減少量は u の増加量に等しい。u に関しては生起率は一定で、単位質量の生起率を q とすると、q*u*du が u に関する生起率積分の減少量である。x に関しては、生起率積分の増加量は R_x*dx である。よって、

(15)　　　$R_x = \tau^2/(G_1 * x^2)$

安定性理論

xに関しては無限大からxまで積分し、uに関しては0からuまで積分すると、

$$\tau^2/(G_1 * x) = qu^2$$

よってu^2はxに反比例している。

(16) $\quad u^2 = q_x/x \qquad q_x = \tau^2/(G_1 * q)$

上式（16）の如く関数（u）を定義すると、質点（A）は（u）を確率変数として正規分布を成す。変数（x）が無限大の時に最も安定、すなわち存在の確度が最も高くなる。関数（u）については$u = 0$の時に存在の確度が最大となる。平均値に関しても、如何なるuの値に関しても生起率が一定であれば平均値は0である。よって、質点（A）の偏差の確度分布は次式を成す。

(17) $\quad \Phi = \exp[-u^2/(2\eta^2)]/[\eta(2\pi)^{1/2}]$

ηは標準偏差値である。この式に上式（16）を代入すると次式を得る。

(18) $\quad \Phi = \exp[-1/(2x\eta^2)]/[\eta(2\pi)^{1/2}]$

これがx軸上の存在の確度分布である。持続の確度分布は揺動0を中心とした正規分布であり、式（7）に従う。

(7) $\quad \Psi = \exp[-V^2/(2\tau^2)]/[\tau(2\pi)^{1/2}]$

この2式に安定性理論の存続性を応用すると、

(19) $\quad d^2x/(dt)^2 = -(\tau^2/\eta^2)/x^2$

となり、原点に距離の二乗に反比例した加速度を与える。

式（14）から、$x = 1$の時、

$$R_1 = \tau^2/G_1$$

式（5）から、

$$\eta^2 = R_1/E$$

$$\eta^2 = \tau^2/(G_1*E)$$

上式を式（19）に代入する。また、質量にするため $E = mC^2$ を代入すると、

$$(20) \quad -d^2x/dt^2 = (G_1*E)/(x^2)$$

よって、点 $(x, 0)$ に存在する質点はその質量（m）に比例し、距離（x）の二乗に反比例する求心力を生じる。

4．旋回

　式（19）は存続性による負の加速度の存在を示す。式（19）の揺動性と散在性の比は角速度であるから点（a）の質点は旋回することを示す。

　また、式（18）と式（7）に変動性を適用すると、次式を得る。

$$(21) \quad d^2x/dt^2 = (\tau^2/\eta^2)/x^2$$

正の加速度は遠心力を生じる。存続性は式（19）が示す如く求心力を生じるので質点は回転中心との距離を変えずに旋回する。

安定性理論

G. 基本法則

1. 散在性と引力

単位エネルギーが原点に存在すると、その近傍に散在性（σ_1）の存在確率分布、$\exp[-X^2/(2\sigma_1^2)]/[\sigma_1(2\pi)^{1/2}]$ をもたらし、その偏位 X 上にエネルギー K の質点が存在した場合、質点 K は存在確率に基づき $X=0$ への求心力（$-F$）を得る。単位エネルギー系がエネルギー K に X の位置でもたらす求心力は、

$$(22) \quad -F = K\exp[-X^2/(2\sigma_1^2)]/[\sigma_1(2\pi)^{1/2}]$$

原点に存在するエネルギー頻度がこの求心力を増加させる。原点にエネルギー E があれば点 X に存在するエネルギー K の存在確率分布が原点のエネルギー E の存在頻度を規定する。それは $E\exp[-(-X)^2/(2\sigma^2)]/[\sigma(2\pi)^{1/2}]$ であり、点 X に存在するエネルギー K の求心力を増加させる。

このエネルギー頻度と式 (22) の積が求心力（$-F$）となる。

$$-F = E\exp[-(-X)^2/(2\sigma^2)]/[\sigma(2\pi)^{1/2}]$$

$$*K\exp[-X^2/(2\sigma_1^2)]/[\sigma_1(2\pi)^{1/2}]$$

$$= KE\exp[-(-X)^2/(2\sigma^2)-X^2/(2\sigma_1^2)]/(2\pi\sigma_1\sigma)$$

$$= KE\exp[-X^2(\sigma^2+\sigma_1^2)/(2\sigma^2\sigma_1^2)]/(2\pi\sigma_1\sigma)$$

X の値が十分に大きい場合はテイラー展開を応用して近似式にする。

$$(23) \quad -F = KE[2\sigma^2\sigma_1^2/X^2(\sigma^2+\sigma_1^2)]/(2\pi\sigma_1\sigma)^2$$

$$= KE[\sigma\sigma_1/(\sigma^2+\sigma_1^2)]/(\pi X^2)$$

ニュートンの万有引力の式を得る。

$$-F = KE\{\sigma\sigma_1/[\pi(\sigma^2+\sigma_1^2)X^2]\}$$

(24) $\quad -F = G*KE/X^2$

Gは引力定数であり、$G = \{\sigma\sigma_1/[\pi(\sigma^2+\sigma_1^2)]\}$ である。

2．クーロン力

　単位エネルギーの山があれば、それは脆く、その周囲に遠心力が発生する。エネルギーの山Eが偏位Xにあれば、それを＋のクーロンの塊と考えるとX＝0と反対方向の力、遠心力を持つ。エネルギー塊中心部のエネルギー頻度は$E/[\sigma(2\pi)^{1/2}]$であり、その遠心力は式（22）を用いて得られる。もし存在確度分布の中心部（X＝0）にKのエネルギーがあったら、位置Xに於けるエネルギーEの遠心力にも変化を起こす。この場合、クーロン力は遠心力である。X＝0の位置のエネルギー存在頻度はエネルギーEとエネルギーKの存在頻度の積である。この場合、散在性（σ）は共通であり、エネルギーEのクーロン力（F）は遠心力である。Xが十分大きければ、式（23）の如くテイラー展開を用いて近似値を得る。クーロン力の式が得られる。

$$F = KE/(2\pi X^2)$$

分布の中心（X＝0）にエネルギーの谷（−K）が在ったとしたなら、そのエネルギーは負のクーロンであり求心力となる。

(24)′ $\quad -F = KE/(2\pi X^2)$

負のクーロンは正のクーロンを引き寄せる。エネルギー（E）のクーロン（q）は次式となる。

安定性理論

$$q = E/(2\pi)^{1/2}$$

3．質量とエネルギー

　静止状態の質点もある強度をもって、その総エネルギーを維持している。質点の運動速度が上昇すると総エネルギー（E）も上昇する。質点の運動速度に限界があるなら、速度と同じ次元を持つ揺らぎの速度である揺動性（τ）の大きさにも限界がある。揺らぎの速度も最大速度 C 以上になることはない。

　式（8）の両辺に最大速度（C）を乗じる。

$$CP = CE_1/[\tau(2\pi)^{1/2}]$$

となり、CP はエネルギーの次元を持ち、E_1 の $C/[\tau(2\pi)^{1/2}]$ 倍のエネルギーであり総エネルギー（E）である。

(25)　　　$E = CP$

質点の静止エネルギーを E_0 とし、運動中の質点の総エネルギーを E とする。

　上式より $E_0 = CP_0$ であるから、式（10）は次式となる。

(26)　　　$E = E_0/[1-(v/C)^2]^{1/2}$

上式の近似式は次式となる。

$$E = E_0 * [1+(v/C)^2/2]$$

$$= E_0 + (P_0/C)v^2/2$$

運動エネルギーは $(P_0/C)v^2/2$ で、ニュートンの運動エネルギー（$m_0 v^2/2$）はこれに一致しなければならない。

$$E = E_0 + m_0 v^2/2$$

ゆえに、

$$m_0 = P_0/C$$

となる。

等式（10）の両辺をCで除すれば、

$$P/C = (P_0/C)/[1-(v/C)^2]^{1/2}$$

$m_0 = P_0/C$ であり、$P/C = m$ とすると、m は運動中の質点の質量を表す。

式（25）より $E/C = P$、よって $E/C^2 = m$、等式（26）の両辺を C^2 で除すると、

$$E/C^2 = (E_0/C^2)/[1-(v/C)^2]^{1/2}$$

質点のエネルギーを最大速度の二乗で除したもの（E/C^2）が質量（m）であり、次式を得る。

(27)　　$m = m_0/[1-(v/C)^2]^{1/2}$

ただし、m は質量、m_0 は静止質量

4．事象の安定化

(1) 実在性

　事象は具体化、実態化される。その実在は存続するか変動するかである。事象は状況が変化すれば変動して安定に存続する。変動は安定化のためである。壊れ易い物の変動性は大きく存続性は小さい。また頑丈な物が壊れないのは、変動性が低く存続性が高いからである。事象の実在性（Θ）に於いては変動性（Z）と存続性（H）が反比例していて、次の

式が成り立つ。その比例定数（Θ）が実在性で定数である。

$$(28) \quad Z*H = \Theta$$

両辺の対数をとると、

$$\mathrm{Log}(Z) + \log(H) = \log(\Theta)$$

そして、微分すると、$d\Theta/dt/\Theta = 0$であるから、

$$(29) \quad dZ/dt/Z + dH/dt/H = 0$$

存続性（H）は事象の性質であり時間に関係しないから、$dH/dt/H = 0$。よって変動性も、$dZ/dt/Z = 0$。ゆえにHとZも定数である。

(2) 変動性

　変動とは、持続確度と存在確度が比例して変動することであり、その比例定数を変動性と定義する。事象は変化する。壊れ易い物は変化し易いが存在状況が不安定ならより変化し易い。変動とは存在の安定化で、揺動の確度（強度）と散在の確度（精度）とは増減順方向であり、変動性は揺動性と散在性とが比例関係にある。変動性（Z）は存在確度（Φ）と持続確度（Ψ）との比例定数となる。

　変動性（Z）は、

$$(30) \quad \Phi/\Psi = Z \qquad \text{ただし、}Z\text{は定数}$$

両辺の対数をとり、微分すると、

$$(31) \quad (d\Phi/dt)/\Phi - (d\Psi/dt)/\Psi = (dZ/dt)/Z$$

変動性に基づき存在確度の変化率と持続確度の変化率の差は0となる。x軸上の1点aに存在する質点Aの存在確度（Φ）と持続確度（Ψ）は比例しているのであるから、(2)と(7)の式を上式に代入すると次式

を得る。

$$(32) \quad (dX/dt)X/\sigma^2 - (dV/dt)V/\tau^2 = (dZ/dt)/Z$$

ただし、Xはx軸上の点aからの偏位で、$V = dX/dt$。Zは比例定数であるので次式を得る。

$$(33) \quad d^2X/dt^2/\tau^2 - X/\sigma^2 = 0$$

$\tau/\sigma = \omega$とするとωのディメンジョンはT^{-1}、角速度であるから次式を得る。

$$d^2X/dt^2 = \omega^2 X$$

点Xに存在する質点は回転している。正の加速度が遠心力であり回転を起こしている。質点がx軸に沿って運動している場合を考え、式(33)を次のように展開する。

$$[d/dt + (\tau/\sigma)][d/dt - (\tau/\sigma)]X = 0$$

$[dX/dt - (\tau/\sigma)X = 0]$は拡散するので、上式の解は次式となる。

$$(34) \quad dX/dt + (\tau/\sigma)X = 0$$

また、この式の解は次式の如く減衰指数関数である。

$$(35) \quad X = X_0 * \exp[-(\tau/\sigma)t]$$

X_0は初期偏位値で、偏位はτ/σの減衰係数で0に収束し安定となる。それゆえ、後に出る波動関数(38)の振幅は0に収束し安定となる。

⑶ 存続性

　存続とはある事象が安定して存在していることである。安定な存在である存続を考えると、存在し難い事象も安全な存在状態下であれば存続する。不安定な状況（散在性の高い）でも壊れ難い（強度の高い）物な

ら存続する。安定な存続には存在状況が安全であるか、丈夫な物であるかの必要がある。物体に限らず事象が精度の低い状況に存続するには高い強度が必要であり、強度の低い事象が存続するには安全な精度の高い状況が必要となる。存続性（H）に於いては、揺動確度と散在確度は反比例している、よって、位置（a）に最尤値を有する質点 A の最尤値上での存在確度（Φ）と持続確度（Ψ）は反比例している。それらの積は定数となる。

$$(36) \quad \Phi * \Psi = H$$

両辺の対数をとり、微分すると、

$$(d\Phi/dt)/\Phi + (d\Psi/dt)/\Psi = dH/dt/H$$

(2) と (7) の式を上式に代入すると次式を得る。

$$(dX/dt)X/\sigma^2 + (dV/dt)V/\tau^2 = dH/dt/H$$

X は x 軸上の点 a からの偏位、V = dX/dt で、H は比例定数で次式を得る。

$$(37) \quad d^2X/dt^2/\tau^2 + X/\sigma^2 = 0$$

上式は波動関数である。$\tau/\sigma = \omega$ とするなら、ω のディメンジョンは T^{-1} であり、波動角速度のディメンジョンとなる。また、次式の波動関数を得る。

$$(38) \quad d^2X/dt^2 = -\omega^2 X$$

質点は X に比例する負の加速度を生じて回転している。波動関数としては物質の波動性の説明でありド・ブロイ波の説明となる。その波長（λ）は $\lambda = h/P$ となっている。光は波であると同時に粒子としての性質がある。ならば粒子も波としての性質もあるべきだ。電子の波長は可視

光線の波長の1000分の1以下である。粒子や質点には波動が潜在している。シュレーディンガーの波動関数にも用いられている。実在性の論理は、これらの確立された事実と合致する。

(4) 回転

　質点（A）がある空間内 x 軸上の一点（b）に存在するとする。もし同一 x 軸上で存在確度の高い点（a）があるとする。質点は点（b）から点（a）に x 軸上を移動する。点（b）の位置は安定な位置（a）の偏位（X）と考えることができ求心力が作用している。移動は変動性の式（34）に基づき移動し、存続性の式（37）に基づき静止する。実在性に基づき、これらからの二式の和は安定化の運動を示す。

　変動性の式（34）に基づき質点は点（b）から点（a）への運動を起こす。式（34）を次式の如くに変換する。

$$(39) \quad 2\omega dX/dt + 2\omega^2 X = 0$$

存続性の式（37）もこの運動に寄与する。

$$(37) \quad d^2X/dt^2 + \omega^2 X = 0$$

式（29）に基づき、この運動の実在性は式（39）と式（37）の和であり、次式の如くである。

$$(40) \quad d^2X/dt^2 + 2\omega dX/dt + 3\omega^2 X = 0$$

この式の解は下記の如くである。X は実数でなければならないから $X_2 = 0$。

$$X = \exp(-\omega t) * [X_1 * \cos(2^{1/2}\omega t) + iX_2 * \sin(2^{1/2}\omega t)]$$

$t = 0$ の時、すなわち X の初期値を X_0 とすると $X_1 = X_0$。

　結果として式（41）が得られる。これは減衰波動関数である。

安定性理論

$$(41) \quad X = X_0 \exp(-\omega t)\cos(2^{1/2}\omega t)$$

振幅は次第に減少し0に近づき質点は安定となる。安定化した質点には角速度があるゆえ、点(a)上で回転している。よって式(33)による遠心力が存在し、同時に式(37)による求心力も存在する。双方の均等な力により質点は安定して点(a)上に存続する。

5. 事象の不確定性

(1) 不確定性原理

事象には常に散在性と揺動性がある。ばらつきやふらつきが100％ない事象はないので少しの散在性や揺動性は必ずある。その散在性や揺動性が事象の変動性や存続性をもたらしている。事象は変化したり継続したりする。不確定性原理は一般に $\Delta E * \Delta t \geq h$ または $\Delta P * \Delta x \geq h$ と表示される。この2つの表示法は同一である。エネルギーの距離微分 $\Delta E/\Delta x \fallingdotseq F$ は力であり、モメンタムの時間微分 $\Delta P/\Delta t \fallingdotseq F$ も力である。よって $\Delta E * \Delta t = \Delta P * \Delta x$。双方の表示は同じである。ここではこの原理を説明するのにモメンタムを用いる。モメンタムにも存続性と変動性がある。モメンタムが、P_0 から P に増加している時、その増加カーブは変動性による減衰指数関数である。変動性でハイゼンベルグによる不確定性原理が説明される。

(2) モメンタムの変動性

モメンタム(P)にも散在性や揺動性がある。モメンタムを微分すると従属変数は力となる。よってモメンタム持続確率の統計変数は力である。モメンタムと力が変動性と存続性を構成する。モメンタムの変動性は次のように求められる。

モメンタム(P)の存在確率分布(Φ)は標準偏差を(σ_P)として、

G．基本法則

$(P=P_0)$ の点の周りでは式 (1)′ の如くとなる。

$$(1)′ \quad \Phi = \exp[-(P-P_0)^2/(2\sigma_P^2)]/[\sigma_P(2\pi)^{1/2}]$$

σ_P は Φ の標準偏差値

P_X を (P_0) からの偏位とすると、$P_X = P-P_0$ で、その分布は次式の如くである。

$$(2)′ \quad \Phi = \exp[-P_X^2/(2\sigma_P^2)]/[\sigma_P(2\pi)^{1/2}]$$

揺らぎは偏位 (P_X) の時間的な変化であるから、力の次元である。

$$(6)′ \quad F = dP_X/dt$$

モメンタムの持続確率分布 (Ψ) の変数は力となる。

$$(7)′ \quad \Psi = \exp[-F^2/(2\tau_P^2)]/[\tau_P(2\pi)^{1/2}]$$

変動性に於いては散在性と揺動性が比例している。変動性 (Z) は式 (30) のように存在確度 (Φ) と持続確度 (Ψ) の商であるから式 (2) と式 (7) とを式 (31) に代入し、処理すると次式 (32)′ を得る。

$$(32)′ \quad (dP_X/dt)P/\sigma_P^2 - (dF/dt)F/\tau_P^2 = (dZ/dt)/Z$$

P_X は p 軸上の存在位置 (P_0) からの偏位であり、F は $F = dP_X/dt$。
　変動性 (Z) は定数であるから次式を得る。

$$d^2P_X/dt^2/\tau_P^2 - P_X/\sigma_P^2 = 0$$

$$[d/dt+(\tau_P/\sigma_P)][d/dt-(\tau_P/\sigma_P)]P_X = 0$$

$dP_X/dt-(\tau_P/\sigma_P)P_X = 0$ は拡散するので、上式の解は次式となる。

安定性理論

$$(42) \quad dP_X/dt + \omega_P P_X = 0$$

⑶ 不確定性と変動性

不確定性原理は一般に次式の如く表記される。

$$\Delta x * \Delta P \geq n*h \qquad n \geq 1$$

$\Delta x * \Delta P$ を微分する。

$$d(\Delta x \Delta P)/dt = \Delta x * d\Delta P/dt + \Delta P * d\Delta x/dt$$

Δx と ΔP に変動性を応用すると、式 (34) より $d\Delta x/dt = -\omega_1 \Delta x$、$d\Delta P/dt = -\omega_2 \Delta P$ であるから、

$$(43) \quad d(\Delta x \Delta P)/dt = -\omega_1 \Delta x \Delta P - \omega_2 \Delta x \Delta P$$

$$= -(\omega_1 + \omega_2) \Delta x \Delta P$$

$\omega_3 = \omega_1 + \omega_2$ とすると $\quad = -\omega_3 \Delta x \Delta P$

Δx と ΔP は正の値であり、$\Delta x \Delta P$ は減少関数と考え得るので、$d(\Delta x \Delta P)/dt$ はエネルギーの次元であるが負の値である。エネルギーはエネルギー粒子 $h\omega$ の n 倍であるから、

$$d(\Delta x \Delta P)/dt = -nh\omega$$

$$-\omega_3 \Delta x \Delta P = -nh\omega$$

ω と ω_3 はいずれもエネルギー $d(\Delta x \Delta P)/dt$ の角速度であるから同一である、

$$\Delta x \Delta P = nh \qquad n \geq 1$$

ゆえに、

$\Delta x \Delta P \geqq h$

6．熱と光の特性

　熱もエネルギーが集まっている状態である。よって熱にも散在性（σ）と揺動性がある。熱は基本的なエネルギーがばらばらになって存在している状態と考えられる。熱にも変動性と存続性がある。熱は空間内で大なり小なり局在していて、その存在確度は低いであろうが平らではない。散在性（σ）は有限である。熱はゆっくりとどの方向にも浸潤して簡単に変化するから、その揺動性（τ）は非常に大きい。よって角速度（ω＝τ/σ）も大変大きい。存続性の波動関数式（38）から、エネルギーである熱には大変高い角速度又は角振動数があることがわかる。

　光も空間内に存在する。そのエネルギー波は物質波で4000〜8000Åの波長を有する。ゆえに光量子には散在性と揺動性がある。光はある瞬間には空間内の一点に散在性をもって存在する。光の存在確度はかなり高くて、散在性（σ）は小さくても０ではない。光の揺動性は大変小さい。光の速度は大変速いので式（9）からその揺動性（τ）は０または０に近いものであることがわかる。よって角振動数（ω＝τ/σ）もほとんど０である。式（35）が示す変動性ではほとんど変化はなく、定数に近い。式（38）が示す存続性では波動を示さず、光はまっすぐ進み周りのものから影響をうけない。変動性の式（35）が示すように、その変動性は一瞬のうちに０まで下がってしまう。光はほとんど変動性がなく、空間内を一直線に進行する。光はエネルギー量子から成り立っていて、その波はその一直線内に閉じ込められている。

7．変動性とロレンツ関数

　ロレンツ関数は変動性から導くことができる。次に示す式（33）は変

安定性理論

動性から導かれた式である。

$$(33) \quad d^2X/dt^2/\tau^2 - X/\sigma^2 = 0$$

両辺を積分する。C_0 は時間に関係しない定数である。

次式で、上の式の τ は運動中の時空座標に於ける揺動性の値で、下の式の τ' は基本の時空座標に於ける揺動性の値である。

$$\sigma^2 dX/dt - \tau^2 X^2/2 = C_0$$

$$\sigma^2 dX/dt - \tau'^2 X^2/2 = C_0'$$

上の式から下の式を引く。

$$\tau'^2 - \tau^2 = 2(C_0 - C_0')/X^2 \quad よって$$

$$\tau'^2 = \tau^2 - 2(C_0' - C_0)/X^2$$

τ'^2 と τ^2 の双方のディメンジョンは速度の二乗なので、$2(C_0'-C_0)/X^2$ を $f'v^2$ と置く。f' は正の定数である。

$$\tau'^2 = \tau^2 - f'v^2$$

v が増加するに従い τ'^2 は減少するが、τ'^2 は0または正であるから τ'^2 が0の時、v は最大速度 C となる。

$$f' = \tau^2/C^2$$

$$\tau'/\tau = 1/(1-v^2/C^2)^{-1/2}$$

この式の右辺がロレンツ関数である。

H．特殊相対論とロレンツ関数

1．運動の相対性

　式（10）の右辺の補正項 $1/[\,1-(v/C)^2\,]^{1/2}$ は質量、エネルギー、モメンタム等の補正に必要な関数であり、これを速度に関する γ 補正関数と呼ぶ。γ 補正関数は 1 以上の値となり、座標変換等にも必要になる。

　質点の運動を二つの座標系の相対性から考え直してみる。ある基準点（R）に準拠する空間を座標系（α）と呼ぶ。この座標系（α）の x 軸方向に速度（v）で移動している基準点（R′）に準拠する空間（座標系〈β〉と呼ぶ）内の一点上に存在する質点（A）の座標系（α）から見た揺動の確度分布（Ψ）を考える。質点（A）の座標系（α）上での揺動性（τ′）と座標系（β）上での揺動性（τ）は必ずしも一致しない。質点（A）はβ上では静止しているがα上では運動しているからである。揺動の平均はβ上では 0 であるが、α上では 0 ではなく、運動速度（v）の影響を受ける。揺動の平均値として速度にある関数（f）を乗じた（fv）とする。この分布の分散（V−fv)^2 の平均値であり、それは V^2 の平均値は τ^2 であるので、τ が fv の近似値であれば次式の如くである。

$$(\tau')^2 \fallingdotseq \tau^2-(fv)^2$$

座標系（β）が座標系（α）に対して静止している時は、$v=0$ であり、（τ′）と τ は一致する。v が大きくなるにしたがって（τ′）^{2} は小さくなるが正または 0 でなければならないから、$f^2=\tau^2/C^2$ となり次式を得る。

　　　(44)　　　$\tau' = \tau/[\,1-(v/C)^2\,]^{-1/2}$

よって τ′ = τ/γ で、τ/τ′の比は γ 補正関数に等しい。座標系（α）から見た質点（A）の揺動性（τ′）は座標系（β）上の揺動性（τ）より小さく

なる。揺動性は速度であり座標系（α）から見た速度は遅くなる。

２．距離の相対性

散在性（σ）の相対性についても揺動性と同様な議論が当てはまる。座標系（α）に対して v の速度で x 方向に移動している座標系（β）上に静止している質点（A）として、この場合での座標系（α）上の質点（A）の期待値との偏位 X の散在の確度分布（Φ）を考える。質点（A）の座標系（α）上での散在性と座標系（β）上での散在性は必ずしも一致しない。座標系（α）から見た散在の確度分布は座標系（β）が運動しているという要素が入るからである。速度（v）で運動している座標系（β）上に静止している質点を座標系（α）から見たばらつき、すなわち存在の確度分布の偏差の平均が０ではなく、速度（v）の影響を受け速度方向に変化するからである。偏差の平均は速度（v）にある関数（g）を乗じたものとする。座標系（α）上での標準偏差値が散在性（σ'）で、その二乗が平均分散 $[(σ')^2]$ で、散在の確度分布の分散 $(X-gv)^2$ の平均である。この平均分散は式（２）に於ける静止時の平均分散すなわち座標系（β）上での質点（A）平均分散（$σ^2$）から偏差の平均（gv）の二乗を減じたものとなるから、

$$(σ')^2 ≒ σ^2 - (gv)^2$$

座標系（β）が座標系（α）に対して静止している時は、（σ'）と（σ）は一致する。（v）が大きくなるにしたがって（σ'）は小さくなるが、正または０でなければならない。よって（v）の最大値を C とすると、

$$g^2 = σ^2/C^2$$

（g）は定数となるので次式（45）を得る。

(45)　　　$\sigma' = \sigma/[1-(v/C)^2]^{-1/2}$

よって、$\sigma' = \sigma/\gamma$ である。散在性（σ）のディメンジョンは長さであるため座標系（α）から見た長さは移動中の座標系（β）の長さより短くなる。「運動している座標系の長さは短くなる」

3. 時間の相対性

　一つのユークリット空間の中にも幾つかのユークリット空間を考え得る。各々の空間は一つの基準点と三つの基準の方向から成り立っている。ある基準点が他の基準点に対して運動している場合もある。ガリレオ相対論では各々のユークリット空間内の位置関係は、その空間の位置ベクトルと基準点の位置ベクトルの和で決まる。この場合、ガリレオ相対論では時間には相対性はなく各空間に共通と考えられていた。

　自然の時間すなわち時刻は常に現在であり、これは如何なるユークリット空間に於いても共通である。過去からの経過時間を負の時間、将来への経過時間を正の時間と考えるなら、時間すなわち、ある時刻から次のある時刻までの経過時間は測定の対象となる連続物理変量となる。経過に関する連続時間では過去も現在も将来も同時に考えられなければならない。いま、ここではある時刻からある時刻までの間、すなわち概念的、数理的な時間を時刻差または数理時間と称する。それに対して物理的に測定される時間を自然時間と称する。自然時間（以後時間と称する）は数理時間すなわち時刻差（以後時刻と称する）に対して不確実性を生じる。時間と時刻の差（T）は0ではなく、平均が0で、0を中心にある標準偏差値をもって正規分布を成している。時間軸上の各点は対応する時刻に対して一定の標準偏差値を持った統計的存在である。ある基準体Rに準拠する時間、すなわち座標系（α）の時間軸上の一点と、その点に対応する時刻との差をTとし、その不確実性の標準偏差値をν

とすると、

$$Y = \exp[-T^2/(2v^2)]/(2\pi v^2)^{1/2}$$

vはYの標準偏差

Yは時刻と時間の差の確度分布であり、0である確率が最も高い。vの次元は時間である。他の基準体R′に準拠する座標系（β）にも上式は当てはまる。この基準体R′が基準体Rに対して静止している場合には上式は座標系（α）から見た座標系（β）の時間にも当てはまる。しかし、基準体R′が基準体Rに対して運動している場合は、座標系（α）から見た座標系（β）の時間の不確実性は上昇する。基準体R′が基準体Rに対して、ある一方向、例えば座標系（α）のx軸方向に速度vで運動している場合の時間と時刻の差（T′）はTに一致しない。T′の平均値は速度の方向でずれる可能性がある。その平均値を速度（v）とある関数（h）との積とする。T′の分散はTの分散とは平均値以外変化はない。よって（T′$-hv$）²の平均はTの平均分散（v^2）に一致する。T′の平均分散は（v'）であるので次式を得る。

$$v^2 ≒ (v')^2 - (hv)^2$$

$v = 0$の時は双方の時間の不確実性は一致する。v^2は正または0であり$v = 0$の時はvが最大で値をCとするとhの値は定数となり、また$h^2 = (v')^2/C^2$となるゆえ、

$$v^2 = (v')^2 * (1 - v^2/C^2)$$

$$(46) \qquad v' = v * [1 - (v/C)^2]^{-1/2}$$

基準点R′が基準体Rに対して運動している場合は、$v' = v * \gamma$であり、基準体Rに準拠する時間に対して基準体R′に準拠する時間は長くな

る。「運動している時間は延長する」よって、各座標系に準拠する時間は必ずしも一致しない。個々固有の時間軸が存在する。

4．座標変換

基準体 R に準拠する座標系（α）の x 方向に速度 v で運動している他の基準体 R′ は準拠する平行な座標系（β）との座標変換を考える。

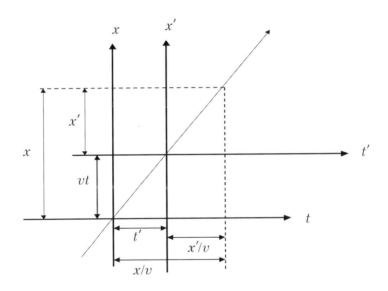

便宜上 x 軸方向のみの変換を考える。基準体 R 及び R′ はある時刻に重なっていたとする。そこから t 秒後には基準体 R′ は x 軸上を vt だけ進む。この時 x が x' と重なっているとすると、上図の x' は $x-vt$ と等しくなる。ただし x' は座標系（α）からは γ だけ短く見えるから、

$$x'/\gamma = x-vt$$

ゆえに、

安定性理論

$$(47) \quad x' = (x-vt)*\gamma$$

運動している基準体に準拠する座標系の距離の変換には速度、時間だけでなく γ 値が関与する。また、時間軸の方を見ると、t' と x'/v の和は x/v に等しい。ただし、運動している座標系の時間は γ だけ長く見えるので、

$$(t'+x'/v)\gamma = x/v$$

$$vt' = x/\gamma - x'$$

$$vt' = x/\gamma - (x-vt)\gamma$$

$$vt' = [vt - x(1-1/\gamma^2)]\gamma$$

$$vt' = (vt - v^2 x/C^2)\gamma$$

$$(48) \quad t' = (t - vx/C^2)\gamma$$

運動している基準体に準拠する座標系の時間の変換には距離、速度だけでなく γ 値が関与する。

エピローグ

　そもそも古来の人間の考える概念は、神の摂理により自然が、またはその一部が創造されたとする。神の摂理である自然は完全で絶対で不安定性はないと考える。しかし、自然は、その全部またはその大部分は偶然により創造されてきたのである。その偶然性のもたらす法則に基づいている。その原理は安定性であり、安定性理論を用いて物理的基本法則を説明することができる。そもそもこの安定性をもたらす原理は筆者が細胞培養や腫瘍増殖の研究データを解析している時に、成長カーブに波動関数が含まれているのではないかと気付いたが、議論の場では常に測定誤差として処理されていた。しかし、細胞培養の初期にオーバーシュートするのは波動関数の関与で説明できるので、長年波動関数の関与を信じて研究中、細胞の安定性を正規関数で表すと説明できることが分かった。存在状況が変化すると、物は変化して安定化する。存在し易い安定な存在状態に変動する。

　筆者は30年ほど前に、この原理で引力の存在を説明できることが分かり、その後は力学に応用して研究を続けていたのである。引力の成因は散在性にある。特殊相対性の成因は揺動性にある。この二つの素因が変動性と存続性をもたらすことを認識し、事象の実在は変動性と存続性が反比例して存在していると認識するべきである。

　物の安定性について次のことが言える。

　　「壊れ易い物は存在し難いが、適切な条件下では存在する。」

　これは、存続性の低い事象は変動性が高く変化するが、存在状況が変化し変動性の低い状況を得れば存続性は高まるということである。変動性と存続性を分析することにより上記の二つの原理は導かれたのであ

る。この安定性理論を用いた偶然性に基づく統計力学は物理的基本法則を説明することができ、これらの問題に答えてくれた。この原理は多くの事象に応用可能であり、諸種の研究に利用されたい。

赤沼　篤夫（あかぬま　あつお）

1964年3月東京大学医学部医学科を卒業後、放射線医学教室に入り放射線の研究を始める。1968〜1973年まで米国とドイツに留学し、エッセン大学助教やストーニイブルック大学医学部講師を兼務。ブルックヘブン国立研究所で客員研究員を併任し、陽子線の線量分布の研究を行う。1973年帰国し東京大学講師。1978年には米国ロスアラモス国立研究所でパイオンの医学利用の研究に携わる。1985年東京大学準教授に。重粒子線の研究のため1991年放射線医学総合研究所に転勤したが重粒子の研究が上手く進まなかったため1997年退官。現在に至る。

安定性理論
The Theory of Stability

2018年4月18日　初版第1刷発行

著　者　赤沼　篤夫
発行者　中田　典昭
発行所　東京図書出版
発売元　株式会社 リフレ出版
　　　　〒113-0021　東京都文京区本駒込3-10-4
　　　　電話 (03)3823-9171　FAX 0120-41-8080
印　刷　株式会社 ブレイン

© Atsuo Akanuma
ISBN978-4-86641-118-7 C3042
Printed in Japan 2018
落丁・乱丁はお取替えいたします。

ご意見、ご感想をお寄せ下さい。

［宛先］〒113-0021　東京都文京区本駒込3-10-4
　　　　東京図書出版